STRATEGIC EXTENSION CAMPAIGN

A participatory-oriented method of agricultural extension

Ronny Adhikarya, Ph.D.
Agricultural Education and Extension Service (ESHE)
Human Resources, Institutions and Agrarian Reform Division

Food
and
Agriculture
Organization
of
the
United
Nations

A case-study of FAO's experiences

Rome, Italy, 1994

M-67

ISBN 92-5-103570-9 /

Food and Agriculture Organization of the United Nations

PREFACE

One of the means of increasing the effectiveness and efficiency of agricultural extension programmes is through the application of improved and innovative extension methods. FAO's field experiences in the last decade have pointed to the need for extension programmes to be more strategically planned, needs-based, participatory, and problem-solving oriented. Extension planning, field implementation and management processes need to be more systematic, interactive, and holistic. Due to the limited resources available to national agricultural extension services in many developing countries, cost-effectiveness and cost-efficiency are two very important factors to be considered in planning and implementing agricultural extension programmes. Hence, the importance of extension methodologies which rely on strategic planning applications and participatory approaches which can minimize inputs or resources and maximize outputs or results.

This publication on *Strategic Extension Campaign : A Participatory-Oriented Method of Agricultural Extension* describes and summarizes FAO's experiences in developing and applying an extension method which addresses the issues mentioned above. It also discusses both the conceptual elements as well as practical aspects that are relevant for applying the Strategic Extension Campaign (SEC) methodology. The publication includes real examples from various FAO field projects which have utilized SEC principles and techniques in support of their agricultural extension activities. While the method described in this publication focuses on an extension campaign, its concepts and principles can easily be applied to the process of overall extension programme planning, implementation and management. Thus, the thrust of this book is on the process, methodology, and application aspects of SEC.

This publication may be useful to agricultural extension, education and training personnel, especially those who are involved in extension strategy development and programme planning. The SEC processes, principles, and techniques described in this book may also be included in extension training courses or programmes. The lessons learned from actual field experiences of SEC implementations can also be useful for adapting and replicating the SEC process and methods for various agricultural extension programmes in different countries.

This publication was prepared by Dr. Ronny Adhikarya who is the Extension Education and Training Methodology Specialist at the Agricultural Education and Extension Service (ESHE), FAO in Rome, Italy. His useful contributions in writing this publication, and also his technical leadership and initiatives in developing, applying and testing the SEC method, as well as training SEC resource persons, in many countries are hereby acknowledged and greatly appreciated.

T. E. Contado
Chief,
Agricultural Education
and Extension Service (ESHE)

FAO of the United Nations
Rome, Italy

TABLE OF CONTENTS

Preface

---

chapter 1   **page 5**

Introduction

1.1.	**Strategic Extension Campaign: What and Why?**	**5**
1.2.	**The Usefulness of SEC**	**6**
1.2.1	Participatory Planning Approach	7
1.2.2	Needs-Based and Demand-Driven Oriented	8
1.2.3	Strategic Planning and Integrated Systems Approach	8
1.2.4	Human and Behavioural Dimensions	9
1.2.5	Problem-Solving Orientation	9
1.2.6	Cost-Effective Multi-Media Approach	10
1.2.7	Specific Extension Support Materials and Training	11
1.2.8	Built-In Process Documentation and Evaluation Procedures	12
1.2.9	Applicability to Other Extension Programmes	13
1.3.	**Purpose of this Case-Study**	**14**

---

chapter 2   **page 15**

The SEC Process & Context

2.1.	**SEC Operationally Defined**	**15**
2.1.1	Integrated Process and Systems-Approach	17
2.1.2	Staff Training as an Integrated Part of SEC	20
2.1.3	KAP Surveys, Evaluation Studies and Follow-Up Actions	20
2.1.4	Participatory-Oriented Method of Planning, Implementation, and Management	21

2.2. **Suggested Conceptual Framework for Strategic Planning of Extension Campaigns**   **25**

2.3. **Phases in Campaign Planning**   **28**

---

chapter 3   page 65

SEC Methodology Applications

3.1. **SEC Implementation Through FAO Projects**   **65**

3.1.1   Examples of SEC Replications and Implementation Arrangements   66

3.2. **Institutionalization Problems: Neglect of SEC Training**   **68**

3.3. **Planned Replications of SEC: Building Cadres of SEC Planners & Trainers**   **69**

3.4. **Facilitating the Multiplier Effects**   **71**

3.4.1   Training of SEC "Master Trainers"   71

3.4.2   Spreading the "SEC-Fever" Beyond Asia   71

3.5. **Strategy of SEC Implementation at Country Level**   **77**

3.6. **Facilitating SEC Institutionalization and Quality Control**   **79**

3.6.1   Develop SEC Training Curricula and Materials   79

3.6.2   Develop and Utilize a Core-Group of SEC Resource-Persons   80

---

chapter 4   page 81

SEC Programmes: Highlights of Results

4.1. **Bangladesh: the Rodent Control Campaigns**   **85**

4.1.1   Evaluation Methods   90

4.1.2   Evaluation Results   92

4.1.3   Costs and Benefits of the Campaigns   96

4.2. **Malaysia: the Rat Control Campaign**   **101**

4.2.1   The SEC Process   102

4.2.2   Evaluation of SEC Results   119

4.2.3   Costs and Benefits of SEC   124

4.2.4   Sustainability of SEC in Malaysia   127

4.3.   Thailand: the Pest Surveillance
System Campaign   131

4.4.   Malaysia: the Integrated Weed Management
Campaign   151

4.5.   Zambia: Assisting Small Farmers
Through Maize Production Campaign   173

4.5.1   Staff Development   173

4.5.2   Applying SEC for Extension Programme Planning   174

4.5.3   Project Evaluation   175

4.6.   Applying SEC in Population Education
Through Agriculture Extension Programmes   181

4.6.1   PEDAEX Results   181

4.6.2   Future Directions for PEDAEX Replications   188

4.6.3   Lessons Learned from PEDAEX Activities   188

chapter 5   page 195

Lessons Learned: Usefulness of SEC for Improving Extension System and Programmes

5.1.   SEC Enhances the Agricultural
Extension Planning Process   195

5.2.   SEC Builds Cadres of Extension
Programme Planners and Trainers   197

5.3.   SEC Helps in Improving
Extension Linkage with Research   198

5.4.   SEC is Needed Most by Small,
Resource-Poor Farmers   199

5.5.   SEC Helps in Improving
Extension Linkage with Training   200

5.6.   SEC Reduces Extension System's
Workload and Increases Its Coverage   201

4

5.7.   SEC Encourages Partnership with,
        and Participation of,
        Community-Based Organizations   203

5.8.   SEC Helps Revitalize Extension Workers'
        Professionalism   204

5.9.   SEC Shows that Extension Programmes
        Can be Strategically Planned,
        Efficiently Managed, and Systematically
        Monitored & Evaluated   205

5.10.   SEC Can Contribute in Improving and
        Strengthening Agricultural Extension
        Systems and Programmes   206

References   **210**

1. INTRODUCTION

1.1.
Strategic Extension
Campaigns:
What and Why?

A "strategic extension campaign" (SEC) methodology developed by FAO has been introduced in Africa, the Near East, Asia and Latin America. This methodology emphasizes the importance of **people's participation** (i.e., intended beneficiaries such as fields extension workers and small farmers) in strategic planning, systematic management, and field implementation of agricultural extension and training programmes (see Fig. 2-4). Its extension strategies and messages are specifically developed and tailored based on the results of a **participatory** problem identification process on the causes or reasons of farmers' **non-**adoption, or inappropriate practices, of a given recommended agricultural technology or innovation. The SEC technology transfer and application approach is needs-based, demand-driven, and has a problem-solving orientation.

The SEC programme follows a systems-approach which starts with farmers' Knowledge, Attitude and Practice (KAP) survey whose results are used as planning inputs and benchmark/baseline for summative evaluation purposes. In addition, a series of practical and participatory approach workshops are conducted to train extension personnel, subject-matter specialists, trainers and farmer leaders together on the skills of extension programme planning, strategy development, message design & positioning, multi-media materials development, pretesting and production, as well as management planning, implementation, monitoring, and evaluation. One of the strengths of this approach is in orienting and training relevant extension personnel to **apply a systematic, rational, and pragmatic approach to planning, implementing, managing, monitoring and evaluating regular/routine programmes** of an agricultural extension service.

Empirical evaluation studies (using information recall and impact surveys, focus group interviews and management monitoring surveys) of strategic extension campaign methods applied to specific FAO-supported extension activities conducted, for instance, in Bangladesh and Malaysia (on rat control), Thailand (on pest surveillance system), Malaysia (on weed management), Zambia (on maize production), Malawi, Jamaica, and Morocco (on population education), etc. reported positive changes in farmers' knowledge, attitudes and practices vis-a-vis the recommended technologies as well as significant economic benefits.

This SEC method has been replicated with FAO assistance in many countries in Asia, Africa, the Near East and the Caribbean, with topics ranging from line sowing method of rice cultivation, maize production, cocoa cultivation, tick-borne disease control, contour tillage, population education, ploughing with drought animal power, etc. In addition to various SEC replications within a country, the multiplier effects of its method are felt beyond national boundaries. For example, extension specialists from Ghana, Malawi, Ethiopia, France, Malaysia, Thailand, the Philippines, etc. who had been trained by FAO on this SEC method and implemented such programmes have now served as consultants/resource persons to train their counterparts, and/or assisted in similar SEC replications, in Sri Lanka, Philippines, Malaysia, Thailand, China, Liberia, Zambia, Malawi, Kenya, Uganda, Morocco, Tunisia, Rwanda, Burundi, Guinea, Jamaica, Honduras, etc.

1.2.
The Usefulness of SEC

The Strategic Extension Campaign (SEC) is **not** an alternative to the conventional extension programme or activity. SEC is, and should be, an integral part of the programmes of an agricultural extension service. The effectiveness and efficiency of such a service could be increased due to SEC's emphasis on its problem-solving orientation, participatory planning approach, intensive extension personnel training, multi-media materials development, and extension management, monitoring and evaluation procedures. Its activities should be carried out by extension personnel and to support the Ministry of Agriculture's policies, strategies and priority programmes. The strategic extension campaign is useful and important to an agricultural extension service due to the following :

1 It Advocates a Participatory Planning Approach
2 It is Needs-Based and Demand-Driven Oriented
3 It Uses Strategic Planning and Integrated Systems Approach
4 It Considers the Human and Behavioural Dimensions
5 It has a Problem-Solving Orientation
6 It Employs a Cost-Effective Multi-Media Approach
7 It Provides Specific Extension Support Materials and Training
8 It has Built-In Process Documentation
   and Evaluation Procedures
9 Its Method is Applicable to Other Extension Programmes

1.2.1. *Participatory Planning Approach*

This participatory approach extension method is responsive to intended beneficiaries' agricultural development problems and information needs because its extension objectives, strategies, methods, messages, and multi-media materials are specifically developed based on survey results of their knowledge, attitude and practice (KAP) vis-a-vis the recommended agricultural technologies. Such a participatory approach in planning SEC activities increases the degree of relevance, and thus acceptability, of extension messages or recommendations among intended **beneficiaries who are consulted** during the planning process regarding their priority concerns and needs. It does not assume the target beneficiaries (i.e. farmers) to be ignorant or requiring all the information there is to know. Rather, it tries to understand and assess farmers' local indigenous knowledge, values and belief systems on farming practices which may be good, need to be improved, or perhaps need to be discouraged. In short, it follows the well-known principles of rural reconstruction: "start with what people already know", and "build on what they already have".

The Usefulness of SEC

1.2.2. *Needs-Based and Demand-Driven Oriented*

In order to make the best use of available extension resources, SEC activities concentrate on meeting the information, education and training needs of intended target beneficiaries. Rather than providing them with the spectrum of information and skills related to a given recommended technology, SEC activities are geared at **narrowing the gaps** between knowledge, attitudes, and/or appropriate practice levels of the target beneficiaries vis-a-vis the technology recommendations. Furthermore, the focus of the SEC activities is to create a demand (through information and motivation approaches) and/or to satisfy the demand (through education and training) among the intended target beneficiaries for the necessary relevant knowledge and skills for adopting the recommended technologies. Such a method needs to apply bottom-up and participatory planning procedures which will give **high priority in meeting the interests and needs of the target beneficiaries.** Tailoring the SEC messages and activities to the specific needs of the intended beneficiaries would not only increase the chances of success but also would increase the efficiency in resources utilization.

1.2.3. *Strategic Planning and Integrated Systems Approach*

The SEC method advocates an integrated and holistic approach in extension strategy development, programme planning and management, training, media and/or materials development, and monitoring & evaluation. To ensure its relevance to audience needs, and to utilize its resources efficiently, it relies heavily on both quantitative data and qualitative information obtained from target beneficiaries (i.e., farmers) to assist in problem analysis, objective formulation, strategy development, and management planning. It applies a **strategic planning** approach in programming and managing its activities, to achieve **maximum** outputs or results using **minimal** inputs or resources in the **shortest time** possible. SEC activities such as surveys, strategy and management planning, multi-media materials design and development, training, field implementation, monitoring and evaluation are integrated as a system which is also an integrated part of a larger extension programme which has linkages with relevant agencies/units dealing with research, inputs/supplies, training, marketing, etc.

1.2.4. *Human and Behavioural Dimensions*

In order to minimize heavy "technology-bias" of many extension activities, the SEC method gives adequate considerations to **human behavioural** aspects, such as socio-psychological, socio-cultural, and socio-economic factors which may facilitate or impede adoption, or continued practice by farmers of recommended technologies. Without sufficient understanding of their positive or negative attitudes and behaviour towards a given technology, the "technology transfer" process would be slow and ineffective, especially if the extension emphasis is on appropriate "technology application" by farmers. There is considerable evidence to suggest that non-adoption of a recommended agricultural technology or innovation is often related to, or caused by, **non-technological** factors, such as social, psychological, cultural and economic problems.

The SEC method gives due attention to human and its environmental factors which may influence important decision-making process related to agricultural technology adoption and practices. It employs a behavioural science analysis, based on a participatory needs assessments and problem identification of the target audience, in developing appropriate strategies and tactics to overcome or minimize human-related constraints affecting the agricultural technology transfer and application process.

1.2.5. *Problem-Solving Orientation*

The SEC is particularly distinguished in that it normally focuses on **specific** issues related to a given agricultural technology recommendation. Its main aim is to solve or minimize problems which caused **non-**adoption of such a recommendation by intended target beneficiaries (e.g., farmers). Unlike more conven-

tional extension programmes or activities, it does not "extend" the whole gamut of information on the recommended technology package. Instead, it **selects,** prioritizes, and utilizes only the most relevant and necessary information or facts which can maximize the effectiveness of extension efforts to minimize or solve the identified problems of non-adoption of a recommended technology. It stresses on the need to provide strategic, critical and/or "quality" (rather than large "quantity" of) information, which must also include non-technological information, given that the reasons for non-adoption of agricultural technologies are often related to socio-psychological, socio-cultural and socio-economic factors. Appropriate human behavioural science principles are thus applied to extension problem-solving and in information positioning and utilization, which is **responsive** rather than prescriptive in nature.

The **segmentation** or classification of extension problems, objectives, strategies and information needs according to a target audience's levels of knowledge, attitude and practice (KAP) in regards to a given recommended technology is not only conceptually important, but practical and useful as well. Problems related to low knowledge level require different solutions than those related to attitudinal problems. Similarly, strategies for changing negative attitudes on a recommended technology are likely to be different than those for solving incorrect practices in technology application or convincing people to try and practice a recommended technology. The implications of the different KAP levels would greatly influence the development of problem-solving strategies, message design, selection of multi-media mix (including when and how to utilize group and interpersonal communication channels, such as extension workers), and materials development, as shown in Fig. 2-5. Application of a behavioural modification approach using information based on the KAP levels of the target audience for message development, media selection and materials development alone could significantly increase the cost-effectiveness of extension activities.

1.2.6.   *Cost-Effective Multi-Media Approach*

One of the most common problems or constraints of a national extension service is the shortage of field extension personnel to reach large number of farmers in widely spread geographical areas with inadequate transporta-

tion facilities. Moreover, extension workers are usually overburdened with unneces-
sarily heavy work-load which includes almost everything that has to do with
farmers at the village level. Such an over-reliance on extension workers is neither
technically sound nor operationally efficient. Some extension functions for certain
purposes such as awareness creation, information delivery, motivational campaigns,
etc., can be more effectively and efficiently performed by other means, channels, or
non-extension groups, under the coordination and supervision of extension workers.

Extension workers' work-load could be reduced by mobiliz-
ing appropriate rural and community-based resources, including the increasingly accessible and low-cost mass communication channels (i.e., local radio stations, rural
press, folk/traditional media, posters, flipcharts, silk-screened printed materials,
audio-cassettes, slide-tape presentations, leaflets, comics, etc.) to disseminate standardized and packaged extension messages, as well as in utilizing local volunteers
(such as school teachers and children, local/religious leaders, etc.) to serve as "intermediaries" in reaching farmers. Such an approach does not imply that extension
workers can or will be substituted by these community-resources. Rather, it is a rational approach of using available resources more effectively and efficiently for certain tasks, such as the need to use extension workers for educational or instructional
purposes which requires two-way interactions, field demonstrations, group discussion, etc., which cannot be done as effectively by mass communication channels.

This SEC method employs a multi-media approach
whereby cost-effective **combination of mass, personal and group communication** channels (including extension workers & trainers) and materials are efficiently
utilized to reduce extension cost and efforts, and to increase its effectiveness in
dealing with a larger number of target audience more rapidly.

1.2.7 *Specific Extension Support Materials and Training*

Most extension services in developing countries suffer
from a lack of relevant and practical extension and training materials to support

field activities of their extension workers. Many extension workers rely primarily on their inter-personal communication skills, and thus their time and/or presence during farmers' meetings may not be utilized as effectively and efficiently as it should be.

Providing specifically designed and relevant extension/training support materials to extension workers will not only facilitate their tasks and reduce their heavy work-load. It will also ensure a certain degree of **quality control** in the delivery of technical information or extension message contents. Experience has also shown that extension workers' motivation, enthusiasm, confidence, and credibility would increase if they are given relevant and attractive multi-media support materials which they could use to improve the effectiveness of their extension and training work.

In SEC activities, extension workers are provided with **pretested** extension and training support materials whose messages are specially designed and developed on the basis of the extension programme's problem-solving oriented strategy plan. Furthermore, these extension workers are also given **special training** to ensure their understanding of extension strategies, message contents, and management/implementation plan, as well as to when, with whom, and how they should utilize the various multi-media support extension and training materials.

1.2.8   *Built-In Process Documentation and Evaluation Procedures*

The advantage of employing a Knowledge, Attitude, and Practice (KAP) survey, as one of the tools for participatory problem identification and information needs assessment, is not only limited to obtaining specific baseline data and inputs for planning extension strategies and improving its management operations. It also provides a benchmark information/data for the purpose of qualitative evaluation, in terms of changes in the levels of KAP over-time. In addition, the SEC activities have built-in evaluation procedures, in the forms of **formative** evaluation (e.g., pretesting of materials and Management Monitoring Survey) and **summative** evaluation (e.g., Information Recall and Impact Survey), for which data/information from the target beneficiaries is essential. It uses various participatory-approach evaluation methodologies including among others, quantitative survey, focus group interview, pretesting, recall test, content analysis, field monitoring, and/or cost-benefit analysis.

Another important aspect of SEC is that it does not only provide empirical evaluation results, but it also usually includes a step-by-step documentation of its operational process through summary briefs as well as more detailed printed, audio and/or visual reports/presentations. Such a **process documentation** and evaluation results have proved to be instrumental for facilitating SEC replications and in obtaining necessary policy, institutional as well as financial support.

1.2.9   *Applicability to Other Extension Programmes*

Most, if not all, of the important principles and techniques employed in planning, implementing, and managing SEC activities are **applicable** for developing and implementing any extension programme. The SEC's process, operational phases, and implementation steps (Fig. 2-1 and 2-2) are essentially similar to that of a regular (but well-designed) extension programme. SEC could thus be considered as a "microcosm" of a well-planned agricultural extension programme. It may be safe to assume that if SEC activities can be carried out successfully in a campaign context, which has a very short time period, then SEC processes, methods, and techniques, either partially or holistically, can be incorporated effectively into a regular and institutionalized extension programme which has a longer time span.

Given appropriate training on various SEC principles and techniques, and through direct involvement in undertaking a planned SEC programme, trained extension staff could help in applying a systematic, rational, and pragmatic approach to planning, implementing, managing, monitoring and evaluating **regular/routine** programmes of an agricultural extension service. As indicated in a number of SEC evaluation studies, many extension staff who had been trained and implemented SEC activities, have continued to utilize their skills in developing and implementing other institutionalized extension programmes for various agricultural technologies. Some have replicated the complete SEC process, while others applied only certain SEC principles or techniques. These efforts have been appreciated and welcomed by many senior officials of the ministries of agriculture where SEC activities were undertaken, as they could see its concrete outputs, results and impact.

1.3.
Purpose of this Case-Study

In order to replicate the SEC process and method more rapidly and widely, this publication, which is basically a case-study, describes the strategy for replications and institutionalization of SEC in different countries and for various agricultural technologies, as well as results and lessons learned from such experiences. This case-study provides some insights into understanding the technical, operational and management needs & requirements for successful implementation of sustainable SEC activities.

This case-study will not describe or discuss in detail the concepts or techniques utilized in SEC, because comprehensive, in-depth, and step-by-step explanations, including actual examples, of the Strategic Extension Campaign method had been written in a book by Adhikarya with Posamentier (1987) *Motivating Farmers for Action : How Strategic Multi-Media Campaign Can Help* . Thus, this case-study will only give a brief overview of the SEC process & context, and summarize important results of various SEC programmes undertaken with FAO assistance. Its focus will be on the analysis of relevant experiences and lessons learned from replicating the SEC method and in institutionalizing it as an integral part of agricultural extension programme planning and strategy development process.

2. THE SEC PROCESS & CONTEXT

In this section, important aspects of the Strategic Extension Campaign (SEC) process, and its conceptual framework as well as the ten operational phases will be described. Some relevant examples on various actual field experiences will be provided to illustrate the suggested SEC operational phases.

2.1. SEC Operationally Defined

As commonly known, in the context of agricultural extension, a campaign is one of the methods of extension which can reach a large number of target beneficiaries in a short time period. More specifically, the Strategic Extension Campaign is :

> "a strategically planned, problem-solving, and participatory- oriented extension programme, conducted in a relatively short time period, aimed at increasing awareness/knowledge level of an identified target beneficiaries, and altering their attitudes and/or behaviour towards favourable adoption of a given idea or technology, using specifically designed and pretested messages, and cost-effective multi-media materials to support its information, education/training, and communication intervention activities."

In order to complement and improve the programmes of a national agricultural extension service, the Strategic Extension Campaign (SEC) method has given special emphasis on the following aspects :

☆ A strategic extension campaign is purposive, problem-solving, participatory-oriented, and focuses on a specific issue or recommended technology.

☆ Its goals are consistent with, and guided by, the overall agricultural development policies and extension programme objectives.

☆ Campaign objectives are specific and formulated based on intended beneficiaries' felt needs and problems identified through a baseline survey of their Knowledge, Attitude, and Practice (KAP) vis-a-vis the recommended technology.

☆ A specific campaign strategy is developed with the aim of solving problems that caused non-adoption, and/or inappropriate or discontinued practice, of the recommended technology.

☆ A strategic planning approach is applied in the process of target audience segmentation, multi-media selection, message/information positioning & design, and extension/training materials packaging, development & production, with a view of obtaining maximum output/impact with the least or minimum efforts, time, and resources.

☆ Formative evaluation in the form of field pretesting of prototype multi-media campaign materials is conducted before they are mass-produced.

☆ A comprehensive and detailed campaign management planning is an integral and vital part of the SEC process, and it will not only spell out the implementation procedures & requirements, but such a plan will also be used to develop a management information system, including monitoring & supervision procedures.

☆ Special briefing and training for all personnel who are involved in SEC activities must be undertaken to ensure that they understand their specific tasks and responsibilities and have the necessary skills and support materials to perform such tasks effectively.

☆ Process documentation and summative evaluation to assess the progress of implementation and impact of SEC activities are conducted, and whose results are used to improve its on-going performance (through Management Monitoring Survey), and to determine SEC's results and overall effectiveness (through Information Recall & Impact Survey, Focus Group Interviews, etc.), as well as to draw lessons learned from such experiences for future replications.

2.1.1 *Integrated Process and Systems-Approach*

The conceptual framework of the strategic extension campaign (SEC) follows a generic model originally proposed by Adhikarya (1978) and described in detail in several of his other publications*. The SEC programme planning framework and process is outlined in Fig. 2-1, where all ten operational phases include participatory approach activities (see Fig. 2-4) by soliciting relevant **feed-forward** (i.e., inputs or information on needs) and **feedback** (i.e., comments or information on results) from target beneficiaries (i.e., small farmers, etc.). More details on the SEC conceptual framework are given in Section 2.2 of this Chapter.

The SEC method also advocates the need to carry out extension activities in a **systematic, sequential, and process-oriented manner**, rather than on an ad-hoc basis. It is a **planned** extension programme with inter-related activities to be carried out following a management implementation plan by well-trained personnel within a given time schedule. The operational phases as suggested in Fig. 2-1 should not be implemented or conducted in isolation because these are part of an integrated and systemic process which requires them to **reinforce** each other and to produce a **synergic** effect.

The application of such a systems-approach also points out the need to train staff to master the whole extension process, rather than only some elements of the process or part of the activities. Therefore, as can be seen in Fig. 2-2, the suggested procedures for carrying out a strategic extension campaign include training activities (through skills-oriented workshops) related to the operational phases or implementation steps which follow closely the conceptual framework and process.

---

* R. Adhikarya and H. Posamentier, *Motivating Farmers for Action: How Strategic Multi-Media Campaigns Can Help*, Eschborn, Frankfurt: GTZ, 1987, and R. Adhikarya, "Guideline Proposal for a Communication Support Component in Transmigration Project", Rome: FAO/United Nations, Project 6/INS/01/T, 1978.

Conceptual Framework for Extension Campaign Planning: 10 Operational Phases

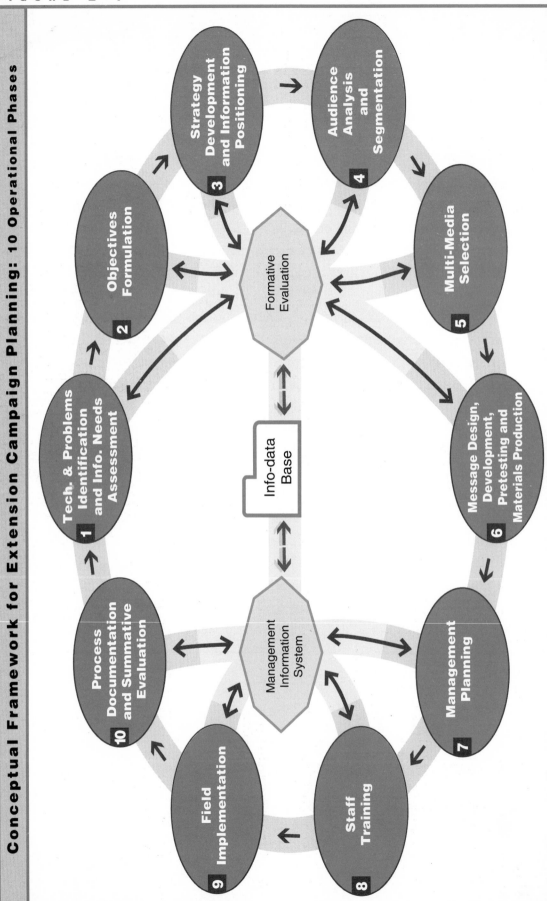

Adapted from: R. Adhikarya and H. Posamentier, *Motivating Farmers for Action: How Strategic Multi-Media Campaigns Can Help*, Eschborn, Frankfurt: GTZ, 1987
For more detailed information on similar frameworks, see: R. Adhikarya and J. Middleton, *Communication Planning at the Institutional Level:
a Selected Annotated Bibliography*, Honolulu: The East West Center, 1979

Implementation Steps for Strategic Extension Campaign & Personnel Training

step	WORKSHOP ON		step	FOLLOW-UP ACTIVITIES
1	Purpose & method of Knowledge, Attitude, and Practice (KAP) survey (10 days)		2	Conduct K.A.P. survey, including Focus Group Interviews (through contractor, max. 3 months)
3	Campaign strategy planning, message design & prototype materials development (14 days)			
4	Methods for pretesting/formative evaluation of prototype multi-media campaign materials (6 days)		5	Field pretesting of prototype campaign materials (0.5-1 month)
			6	Revision and production of multi-media campaign materials (2-6 months)
7	Campaign management planning & training (6 days)		8	Briefing & training of extension workers and other field personnel involved in campaign activities (2-3 months)
			10	Field implementation of campaign (2-6 months)
9	Campaign evaluation methods and management monitoring survey procedures (5 days)		11	Conduct evaluation (max. 4 months) → Management Monitoring Survey (MMS) → Information Recall & Impact Survey (IRIS) → Focus Group Interviews (FGI)
12	Results demonstration & experience sharing on extension campaign planning, implementation & evaluation (5 days)		13	Publication/audio-visual documentation of campaign process, activities & results (max. 3 months)
			14	Replications & improvement of campaign process in other types of technology, locations or countries
			15	Utilization of good trainees as master trainers or resour conduct similar workshops or follow-up activities in oth country or in other countries

2.1.2   *Staff Training as an Integral Part of SEC*

Without a core-group of extension personnel in a country who are well-trained in specific methods and techniques in planning, implementing and managing SEC activities, it would be very difficult to institutionalize SEC into a national agricultural extension service. Such SEC activities could be initiated and conducted with the assistance of international project experts or consultants. However, the SEC's sustainability cannot be ensured unless a reasonable number of extension staff are specifically trained on all important SEC aspects. They must also have adequate and relevant first-hand experience in planning, implementing and managing such activities.

It is, therefore, imperative that in the process of introducing or replicating the SEC method in a country, training of a national group of extension planners, managers and trainers, subject-matter specialists, communication support staff, and field extension officers on relevant SEC methods and techniques be made as an integral part of the SEC programme planning and implementation.

Due to the inclusion of staff training on SEC process and methodology, significantly higher cost and a longer time would be required for the initial implementation of SEC activities. Such an investment on **human resources development** would perhaps be one of the most cost-effective inputs which could significantly contribute towards institutionalization of an SEC approach in the quest of improving and strengthening a national agricultural extension system and service.

2.1.3   *KAP Surveys, Evaluation Studies and Follow-Up Actions*

In addition to having a training component as an integrated part of the extension process, operations research such as baseline/Knowledge, Attitude and Practice (KAP) surveys and other means of formative and summative evaluation must also be **built-in** to the extension process, programme and methodology. The most vital and essential part of the process is the actual field action and implementation activities which will be required for the preparation and/or follow-up of strategic extension campaign training and operations research activities. Based on the conceptual framework presented in Fig. 2-1, we can identify three different but inter-related activity components and implementation steps which are parts of the 10 operational phases. These components as shown in Fig. 2-3, have various types of activities which are conducted in a process-wise sequence as indicated by the various implementation steps (Fig. 2-2).

<table>
<tr><td>2.1.4</td></tr>
</table>

Participatory-Oriented Method of Planning, Implementation, and Management

As mentioned earlier, the Strategic Extension Campaign (SEC) method relies heavily on the participation of all those who are involved in campaign activities, including the campaign staff as well as target beneficiaries, in providing relevant inputs or suggestions throughout the campaign process. By so doing, the SEC activities are more likely to address relevant problems and issues, as well as to provide practical solutions which are based on the needs of target beneficiaries. As suggested below in Fig. 2-4, efforts should be made to **consult** various types of people who are or will be involved in campaign activities in order to seek their inputs during the different phases and/or steps of a campaign process.

It should be noted that not only the target or intended target beneficiaries of a campaign should be consulted and sought for their inputs to ensure the **relevance** and usefulness of the campaign. Potential campaign implementors, such as concerned field staff, trainers, community leaders, etc. should also be involved and given a role to play in different aspects of campaign planning, implementation and management. Various means of obtaining their inputs and support can be used, and regardless of the methods, an important purpose for seeking their active participation is to make them **"share-holders"** of the SEC programme or activities.

It is not practical and realistic to expect that all target beneficiaries and campaign staff would be involved in every aspects of SEC activities.

FIGURE 2-3

Components, Activities, and Implementation Steps in Strategic Extension Campaign (SEC) Operational Process

COMPONENT	ACTIVITY	IMPLEM. STEP (see Fig. 2-2)
1. SURVEYS AND EVALUATION STUDIES	a. Knowledge, Attitude, and Practice (KAP) Survey	2
	b. Pretesting of Prototype Multi-Media Materials	5
	c. Management Monitoring Surveys (MMS)	11
	d. Information Recall & Impact Survey (IRIS)	11
2. STAFF TRAINING	**WORKSHOPS ON :**	
	a. Purpose & Methodology of KAP Survey	1
	b. Strategy Planning, Message Design, and Prototype Materials Development	3
	c. Methods for Pretesting/Formative Evaluation of Prototype Multi-Media Materials	4
	d. Extension Campaign Management Planning & Training	7
	e. Evaluation & Management Monitoring Survey Procedures	9
3. FIELD ACTIVITIES/ FOLLOW-UP ACTIONS	a. Conduct the KAP Survey	2
	b. Conduct Field Pretesting of Prototype Multi-Media Materials	5
	c. Revise & Reproduce Pretested Multi-Media Materials	6
	d. Brief & Train Field Extension Personnel	8
	e. Field Implementation	10
	f. Conduct Process & Summative Evaluation Surveys/Studies	11
	g. Prepare Process & Results Documentation in Printed, Audio and/or Visual Forms	12
	h. Disseminate and Share SEC Results and Experiences	13
	i. Replicate & Improve Extension Process & Methods for Other Technologies	14
	j. Utilize Good Trainees as Master Trainers/ Resource Persons for Future Extension Campaign Activities	15

FIGURE 2-4

Strategic Extension Campaign and its Participatory and Consultative Process or Mechanisms

OPERATIONAL PHASE (see Figure 2.1)	TYPE OF PEOPLE CONSULTED/ INVOLVED*	MEANS FOR PEOPLE'S PARTICIPATION	IMPLEM. STEP (see Fig. 2.2)
1. Technology & problems identification and info. needs assessment	→ Intended/target beneficiaries → Extension & research staff	→ KAP workshop → KAP/baseline survey → Focus Group Interviews (FGI)	1 2 2
2. Objectives formulation	→ Extension planners & trainers → Farmer/community leaders → Relevant/concerned agri. dev. managers/decision-makers	→ Strategy Planning workshop → KAP/baseline Survey → Focus Group Interviews (FGI)	3
3. Strategy development and information positioning	→ Extension planners & trainers → Farmer/community leaders → Relevant/concerned agri. dev. managers/decision-makers	→ Strategy Planning workshop	3
4. Audience analysis and segmentation	→ Extension planners & trainers → Farmer/community leaders	→ KAP/baseline survey & FGI → Strategy Planning workshop	2 3
5. Multi-media selection	→ Extension planners & trainers → Farmer/community leaders → Intended/target beneficiaries	→ KAP/baseline survey & FGI → Strategy Planning workshop	2 3
6. Message design, development, pretesting and materials production	→ Extension planners & trainers → Farmer/community leaders → Intended/target beneficiaries	→ KAP/baseline survey & FGI → Strategy Planning workshop → Pretesting workshop and field work	2 3 4, 5 & 6
7. Management planning	→ Extension planners → Field extension staff	→ Campaign management planning & monitoring workshop	7
8. Training of personnel	→ Extension planners & trainers → Field extension staff	→ All SEC related workshops	1, 3, 4, 7, 8 & 9
9. Field implementation	→ Extension planners & trainers → Farmer/community leaders → Intended/target beneficiaries	→ Evaluation & management monitoring survey workshop → Management monitoring survey (MMS)	9 & 10 10 & 11
10. Process documentation and summative evaluation	→ Extension planners & trainers → Farmer/community leaders → Intended/target beneficiaries → Relevant/concerned agri. dev. managers/decision-makers → Research staff	→ Management monitoring survey (MMS) → Informaton recall & impact survey (IRIS) → Focus Group Interviews (FGI) → Experience sharing & results demonstration seminars/meetings	10 & 11 11 11 12

* As it is not practical to involve all the people, only a representative sample of those concerned are included in the consultative process.

However, through a careful and systematic method of soliciting fairly specific opinions, ideas, suggestions from a sample of concerned groups of people involved in SEC activities, a useful **campaign map** can be drawn to determine its direction, strategy, types of message contents, training needs, field implementation requirements, etc.

A problem-solving and demand-driven strategic extension campaign cannot be properly designed and planned without relevant **"feed-forward"** information from its intended beneficiaries and concerned campaign personnel. It is thus imperative that a participatory-oriented method of planning be applied to ensure a needs-based extension campaign which can realistically be implemented and provide benefits to its intended clientele.

2.2.
Suggested Conceptual Framework for Strategic Planning of Extension Campaigns

The effectiveness of an extension campaign depends very much on its strategy which should be specific, systematic, and well-planned. In developing a campaign plan, the use of a conceptual framework which can guide the planning process in a systematic, rational, and strategic manner should be considered. A good conceptual framework can help in the application of relevant theories, principles or research findings in the planning and development of sound extension strategies and plans.

In this publication, planning is defined as a process of identifying or defining problems, formulating objectives or goals, thinking of ways to accomplish goals and measuring progress towards goal achievements (Middleton and Hsu Lin, 1975). A good extension campaign plan has a strategy which reflects the target beneficiaries' identified problems/needs and the way information, education (i.e., training) and communication will be used in solving such problems or meeting the needs. Such a plan must outline the management actions to be taken in implementing the strategy. Thus, in this context, campaign planning has to include both strategy planning (i.e. what to do) and management planning (i.e. how to make it happen).

Strategic planning can be operationally defined simply as the best possible use of available and/or limited resources (i.e., time, funds, and staff) to achieve the greatest returns or pay-off (i.e., outcome, results, or impact). Strategic planning is also an approach to anticipatory planning in order to reduce or overcome some uncertainties in decision-making process by prioritizing actions or interventions which may produce the most likely positive outcomes or results.

A strategic extension campaign plan should provide specific guidelines and directions in making information, education and communication activities operational. It must be constantly reviewed, especially at the implementation stage. Adaptation or modification of the plan may be required because of specific local conditions/problems or alteration of the policies/objectives which guided the original plan. The plan should thus be flexible and ready for necessary modification as suggested by feedback results/information (e.g. through process and/or formative evaluation, including pretesting) in order to improve the strategy or the management of campaign implementation activities.

The process of developing a strategic extension plan can be divided into two major parts. The first part is the process of strategy development planning. The second part is the process of management planning. To provide a systematic approach in developing a strategic extension campaign plan, a generic conceptual framework (see Adhikarya, 1978 and also Adhikarya with Posementier, 1987) is suggested based on a 10-phase circular model (Fig. 2-1). The suggested process of developing a strategic extension campaign plan is described below, adapted from the of conceptual framework originally proposed by Adhikarya (1978) :

→ **Part I : Campaign Strategy Development Planning**

Phase 1: Technology & problems identification
and information needs assessment

Phase 2: Campaign objectives formulation

Phase 3: Strategy development and information positioning

Phase 4: Audience analysis and segmentation

Phase 5: Multi-media selection

Phase 6: Message design, development, pretesting and
materials production

During a campaign strategy development planning, as much formative evaluation as possible should be included as a built-in component in all the above-mentioned phases, especially in phases 4 to 6. Formative evaluation in this context means the process of testing the suitability, appropriateness or effectiveness of campaign strategy and plan, including its multi-media messages and support materials, preferably before full implementation, in order to ensure good campaign performance or results.

When a plan for a campaign strategy is completed, it must be translated into action. At that stage, the task of a extension campaign planner shifts from strategy development to **management planning.** To transform extension campaign strategies into extension campaign activities, management objectives must be identified clearly to include at least the following elements: what the action is; who is to carry out the action; how the action is to be carried out; how much resources will be needed and how to obtain such resources; when the action is to be accomplished. In addition, management objectives should set a standard for measuring progress and impact of implementation. Thus, the following phases are necessary in campaign management planning :

→ **Part II : Campaign Management Planning**

Phase 7: Management planning
Phase 8: Training of personnel
Phase 9: Field implementation
Phase 10: Process documentation and summative evaluation

These four phases of campaign management planning should be supported by a management information system to provide planners with regular and up-to-date information for at least the basic components of management objective: who will do what and when. There are three kinds of management activities for which such information is needed to make effective decisions: personnel, finance and logistics. It should be noted that management information system is useful so long as it does not create an unnecessary burden on extension and training staff, distracting them from their basic information, motivation, and education tasks (Middleton and Hsu Lin, 1975).

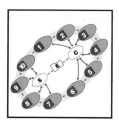

2.3. Phases in Campaign Planning*

Phase 1: **_Technology & problems identification and information needs assessment_**

The planning of a comprehensive extension campaign strategy requires an adequate and accurate set of baseline data including information on recommended technology(ies), identified problems, and information needs. Many types of baseline data need to be collected and analysed. Without such information, a feasible, efficient and cost-effective extension campaign cannot be planned and developed properly. In particular, such information is critical for formulating campaign objectives, developing strategies, selecting and prioritizing extension/educational contents, positioning and designing messages, and evaluating the effectiveness of campaign activities.

Extension planners and practitioners are often forced to make programming or management decisions guided mainly by assumptions or personal experience which might be invalid, and could waste programme resources. Such counter-productive decisions may stem from inadequate planning and/or lack of baseline data and other relevant "feed-forward" information (i.e., data about target beneficiaries' problems, needs or interests on which campaign messages should focus).

---

* This section is largely taken from Chapter 4 of R. Adhikarya's book _Motivating Farmers for Action_ (1987) published by GTZ, Eschborn, Germany. While some of the illustrations in this Chapter are not agriculture-related, they represent clear examples of important processes/principles which can be applied in agriculture extension campaigns.

→ **KAP survey : a tool for participatory strategic planning and evaluation**

As the Strategic Extension Campaign (SEC) method follows a participatory and demand-driven or needs-based approach, target beneficiaries need to be consulted in the process of identifying problems and/or needs regarding their requirements or acceptability of a given technology. A suggested procedure for conducting a participatory assessment of problems and needs is through a baseline survey of target beneficiaries' Knowledge, Attitude, and Practice (KAP) on specific and critical elements of a recommended technology. Unlike other baseline surveys which are often macro in scope and exploratory in orientation, KAP survey is problem-solving oriented and it operates at a micro-level, with a focus on determining at least three distinct conceptual categories : knowledge, attitudes and practice levels of target audiences vis-a-vis the critical elements of a given technology recommendation. The KAP survey also seeks qualitative information from respondents through focus group interviews (FGI), such as on the reasons or causes of their negative attitudes and non-adoption or inappropriate practice with regard to the recommended technologies. Information provided by KAP surveys is very useful, especially for campaign objectives or goals formulation and strategy development as described in the discussion on Phases 2 and 3 of campaign planning.

In many developing countries, extension resources (i.e., funds, facilities, staff, and time) are limited and thus the effective and efficient use of such resources is imperative. A strategic planning approach can help in identifying critical extension/education intervention areas which are important and likely to create a significant impact. Results of a KAP survey can be utilized to analyze which specific elements of the technology package are not known to the majority of target beneficiaries, what are the reasons for their negative attitudes, how and why they have practised recommended technologies inappropriately, etc. Therefore, in most instances, it is unnecessary to provide all target beneficiaries with a complete set of technology recommendations as some of them may already have known, agreed with, and/or acted on, the necessary information. Problem-solving and strategic extension planning basically follows the principle of "start with what they know, and build on what they have".

KAP survey results can also be utilized for audience analysis and segmentation purposes to determine who needs which types of information/messages through what combination of multi-media materials and channels. In addition relevant findings from surveys on media consumption patterns and habits, media availability and reach, and other socio-psychological and anthropological research studies are useful inputs to the exercise of extension strategy planning and message development.

Likewise, a meaningful evaluation of extension campaign activities is dependent on the availability of baseline data such as those obtained through a KAP survey. It is very difficult to make any summative evaluation judgement if there is no basis for comparison (e.g. pre-test vs. post-test measures; control vs. treatment groups). Baseline data can provide the necessary benchmark for comparison of impact or summative evaluation. Even for formative evaluation, availability of baseline data can help in testing and improving campaign process including the soundness of its strategies, and the appropriateness of its messages and multi-media materials as perceived by the target beneficiaries.

Baseline data collection is usually associated with a research activity. The word "research" is often viewed negatively by many programme planners, project managers/administrators, and decision-makers because it is perceived as an undertaking that requires a lot of money, time and specialized personnel. This perception is incorrect as what is needed is an "applied-oriented" or "problem-solving oriented" survey for programme or strategy decision-making rather than pure or experimental research for theory- or hypothesis-testing. A KAP survey is basically a "quick and cheap" way of consulting target beneficiaries and requesting them to provide information on certain trends or an indication of specific directions regarding their problems and information needs rather than elaborate research findings with statistical tests of significance, etc. As described later, most strategic extension campaign programmes which utilized baseline data effectively spent relatively little of their resources or time in acquiring the necessary data or information.

→   **KAP survey : examples of results utilization**

As shown in Fig. 2-5 (see page 40), the audience's KAP level can be used as a basis to consider, in general or broad terms, the type and direction of an extension campaign strategy. While the specifics and details of the strategy still need to be developed further using other types of information and data, the KAP survey is an essential first step in providing information for the logic and rationale for campaign planning and strategy development. For a brief overview on how KAP survey findings have been used in the formulation of specific campaign objectives and in message positioning, design and development, some sample worksheets of a planning exercise of a rat control campaign for the State of Penang, Malaysia, have been included in the following pages. The campaign planning exercise was part of the workshop activities conducted by the author for the FAO Inter-country Programme for Integrated Pest Control in Rice in South and Southeast Asia and the Department of Agriculture, Government of Malaysia, in 1985.

Identified Problems in Rat Control in Penang, Malaysia
Based on findings of a KAP survey of farmers in Penang State

1. Low knowledge of the value of physical methods and cultural practices regarding rat control
2. Low knowledge of different functions and characteristics of different rodenticides
3. Misconception that rats are "intelligent", and thus unlikely to be successfully controlled
4. Lack of group and collaborative efforts to control rats
5. No action to control rats until damages are visible
6. Inappropriate application of rodenticides in different situations
7. Most farmers have more than one job and thus do not have enough time to control rats
8. Superstition that rats would take revenge on behalf of their dead "friends" by causing worse damages
9. Simultaneous planting is not practised, thus providing continuous food supply for rats

---

Source: R. Adhikarya (1985), "Planning and Development of Rat Control Campaign Objectives and Strategies for the State of Penang, Malaysia".
Note: KAP refers to knowledge, attitudes and practice of the target audience.

Specific and Measurable Campaign Objectives for Rat Control Campaign in Penang State, Malaysia

Identified Problems among Farmers	Formulated Extension Campaign Objectives (based on KAP survey results)
1. Inadequate knowledge of the value of physical methods and cultural practices regarding rat control	To raise the proportion of rice farmers' level of knowledge/appreciation concerning the value and benefits of cultural practices from 67% to 75%, and physical rat control practices from 31% to 45%
2. Little knowlege of the different functions and characteristics of different rodenticides	To raise the proportion of rice farmers' level of awareness and knowledge by improving their understanding regarding the different functions and characteristics of two types of rodenticides: a) chronic poison baits from 61% to 70% b) chronic poison dust from 22% to 40%
3. Misconception that rats are "intelligent" and thus unlikely to be successfully controlled	To reduce the proportion of rice farmers' misconception that rats are unlikely to be controlled successfully because they are "intelligent" from 52% to 35%
4. Lack of group and collaborative efforts in controlling rats	To encourage greater participation of rice farmers in group and/or collaborative efforts in controlling rats, by increasing the proportion of rice farmers' level of favourable attitudes towards such efforts from 60% to 70%
5. Farmers normally do not take voluntary action to control rats until crop damages are visible	To increase the proportion of rice farmers who believe that rat control is not a waste time from 55% to 65% in order to encourage them to take action before their crops are damaged
6. Inappropriate application of different rodenticides in different situations/stages	a) To increase the proportion of rice farmers' knowledge on the correct application of rodenticides with regard to: 1. Rate of application of acute poison from 11% to 40 % ; chronic poison (baits) from 23% to 40%; and chronic poison (dust) from 67% to 75% 2. Time of application of acute poison from 47% to 60%; chronic poison (baits) from 39% to 50%; and chronic poison (dust) from 41% to 55% 3. Location to place acute poison from 43% to 55%; chronic poison (baits) from 43% to 55%; and chronic poison (dust) from 78% to 80% b) To increase the proportion of rice farmers' level of appropriate practice in rodenticides application with regard to: 1. Rate of application of acute poison from 12% to 24%; chronic poison (baits) from 23% to 40%; chronic poison (dust) from 32% to 40% 2. Time of application of acute poison from 28% to 35%; chronic poison (baits) from 28% to 35%; chronic poison (dust) from 43% to 50%
7. Lack of motivation of most farmers who have more than one job, to spend more time and effort to control rats in order to increase their yields and income	To motivate and encourage rice farmers to spend more time and efforts to control rats in order to increase their crop yields and income, by increasing their perception that controlling rats is more beneficial than doing other jobs; from 37% to 50%
8. Superstition that rats will take revenge on behalf of their "dead friends" by causing worse damages	To reduce the proportion of rice farmers' misconception regarding their superstitious belief that rats take revenge on behalf of their "dead friends" by causing worse damages from 54% to 50%
9. Non-practice of simultaneous planting which could disrupt food supply for rats during part of the year	To encourage more rice farmers to engage in simultaneous planting in order to reduce time for rats to have continuous food supply by enhancing positive attitudes towards that practice; from 79% to 85%

Source: Adapted from R. Adhikarya (1985)

Example of a Worksheet for:
Extension Planning and
Strategy Development Process Exercise

Problem	Reasons or causes for problem	Problem-solving strategy or approach	Information positioning approach
Farmers' misconception that rats take revenge on behalf of their "dead friends" by causing worse damages	Superstitious belief	As a Moslem, it is sinful to believe in superstitions. Other specific citation from the Holy Book (Quran or Hadith) regarding the above	Religious disincentive
Not enough time to control rats	70% of farmers have more than one job	"Rat control method of using wax and dust poison is simple and easy. Even your wife and children can do it if you are busy".	Task delegation
Inappropriate application of rodenticides at different crop growing stages	Unclear and complicated rodenticide application recommendations	Simplification of technology recommendations and easy-to-remember application procedures	Simplicity
	Inefficent use (too strong a dosage) of rodenticides since they are distributed free to farmers	Arouse guilt feeling of farmers by stressing the waste of their community funds due to inefficient and ineffective use of the free rodenticides	Guilt feeling creation
Farmers' misconception that rats are "intelligent", thus control unlikely to succeed	Failure of zinc phosphide (e.g. bait shyness effect) which is used by the majority of farmers	De-emphasize the use of acute poison (zinc phosphide) and encourage the use of wax poison before booting stage and dust poison after booting stage.	Down-playing the competitor. Easy-to-remember action.
		Stress the need for group and collaborative efforts instead of the individual approach if the battle to fight "smart" rats is to be won	Need for group efforts

Source: Adapted from R. Adhikarya (1985)

Example of a Worksheet for:
Message Design Process Exercise

Problem solving strategy	Message appeals	Examples of message appeals	Channel of message-delivery
Counter-attack farmers' superstitious belief that rats take revenge on behalf of "dead friends" by causing worse damage	Fear arousal

Religious Incentive | "Its is sinful for a Moslem to believe in superstition"

"The more rats you kill, the more you will be rewarded in heaven" (Citation from the Holy Book of Islam) | Religious leaders' sermons during Friday prayers; leaflet; radio spots |
| Discourage the use of zinc phosphide and encourage the wax and dust poison (chronic rodenticides) | Safety, convenient/ simplicity

Testimonials | "Wax poison and dust poison are much safer, easier and more effective than zinc phosphide"

Use testimony from safisfied wax and dust poison adopters about its simplicity of use and effectiveness | Instructional poster; radio spots; pamphlet; portable flipchart; pictorial booklet |
| Motivating farmers to conduct group/collaborative action in controlling rats (e.g. simultaneous planting, applying physical control methods, or conducting simultaneous rat control) | Ridicule

Cultural/ traditional value

Solidarity | "Since rats are 'intelligent' if you fight them alone you might lose and thus be a victim, let's do it together"

"Gotong-Royong (working together and helping each other in a community) is a virtue and the most effective means to control rats"

"Bersatu kita teguh, bercerai kita roboh" (united we stand, divided we fall) | Motivational poster; leaflet; pictorial booklet; slide-sound |
| Educating farmers on the appropriate use/application of rodenticides | Guilt feeling, civic responsibility

Easy to remember | "Don't you feel guilty wasting community funds if you are not using the free rodenticides properly"

"Use wax poison weekly before booting stage and dust poison after booting stage" | Instructional poster; radio spots; group discussions; portable flipchart |

Source: Adapted from R. Adhikarya (1985)

Phase 2:   *Campaign objectives formulation*

In order to be effective, strategic extension campaign should be an integral part of a given agricultural extension system or programme and its function should be to support such extension activities. The specific campaign objectives should reflect the extension system or programme goals, respond to the needs of the programme and its target audience and help in solving the problems encountered in achieving such goals. It should be pointed out, however, that strategic extension campaign objectives are not necessarily the same as the agricultural extension system or programme goals which are expected to be the ultimate results of the whole extension undertaking. The achievement of the campaign objectives is a necessary, but not a sufficient, condition for achieving extension system or programme goals.

An extension programme goal should be explicit in specifying what is to be accomplished. For instance, the following two inadequate goal statements give only the general or operational elements to be achieved :

"To provide irrigation for rural people".
"To drill 4,000 ring wells and 2,000 tube wells by August 1994".

Those descriptions of goals can be made more comprehensive and specific, as well as made to reflect the actual scope of the programme, as follows :

"To increase the number of small farmers in districts X, Y and Z using water from the wells to irrigate their farmland from the present 100,000 to 175,000 small farmers within two years".

With such a goal, the extension programme could not be considered successful if it had only drilled 4,000 new ring wells and 2,000 new tube wells within two years although such an operation or activity was important and essential. It might well be that the water from the wells could not be used for irrigating farmers' land but, perhaps only for household purposes, or not at all. Even when the water might serve irrigation purposes, perhaps it would be mostly the large farmers who were using it rather than small farmers.

The ultimate goal of the above programme or project is not only to drill new wells but to persuade small farmers to utilize water from the wells in irrigating their farmland. Since a campaign activity is an important support component of the extension programme, the chances for success in meeting the programme goal are increased if the campaign objectives of the programme or project are accomplished.

It is thus also useful and important to distinguish extension campaign objectives from the broader extension programme objectives. Strategic extension campaign objectives are usually very specific and aimed at increasing knowledge, influencing attitudes, and/or changing practices (or behaviour) of intended beneficiaries with regard to a particular action recommendation or techonology.

Campaign objectives should specify some important elements or characteristics of the extension activities which could help to provide a clear operational direction, and facilitate a meaningful evaluation of the campaign. Some of those elements are, at least, the following:

1.   the target beneficiaries;
2.   the outcome or behaviour to be observed or measured;
3.   the type and amount/percentage of change from a certain baseline figure expected from the target beneficiaries;
4.   the time-frame;
5.   the location of the target beneficiaries.

It should also be pointed out that if the time frame and/or location of target beneficiaries are not explicitly mentioned in a extension campaign objective, it is usually understood that they follow the time-frame or duration and the target location/area of the campaign itself.

In the case of the above-mentioned irrigation programme, examples of strategic extension campaign objectives which would help in achieving general extension programme goals could include the following:

a   to inform within one year at least 65 percent of the small farmers in X, Y and Z districts about the procedures and benefits of an irrigation system using ring and tube wells;

b   to reduce the proportion of small farmers in districts X, Y and Z who have misunderstandings and misconceptions about the cost and technical requirements of drilling and building ring or tube wells, from the present 54 to 20 percent in one year;

c   to increase the proportion of small farmers in districts X, Y and Z regarding their positive attitude towards the practical and simple use of the irrigation system to water their farmland, from the present 32 to 50 percent within two years.

d   to persuade small farmers in districts X, Y and Z to use water from the wells to irrigate their farmland, and to increase the level of such a practice by them from the present 20 to 35 percent in two years.

It should be noted that campaign objective (d) above is very similar to the stated extension programme goal. While this objective may be considered and included as one of the campaign objectives, it should be realised that the accomplishment of this objective is not entirely dependent on effective campaign strategies or activities. There are other "non-extension" factors which might influence the accomplishment of this objective, such as availability of inputs or funds, type of seeds utilized and land conditions. A campaign is only one of the many factors which can contribute to the successful achievement of a programme or project.

Strategy development and information positioning

A systematic and strategic campaign plan is essential to the efficient achievement of campaign objectives. The first step in a strategic extension campaign (SEC) planning exercise is to identify clearly the problems which may impede or obstruct adoption of the suggested idea, innovation or technology. Data from baseline/KAP surveys, including problem identification and needs assessments results should be analysed carefully. The central issue(s) or problem(s) which might impede progress in achieving the extension goal should be identified. The following two actual cases illustrate the importance of problem identification and analysis :

In a coconut replanting and rehabilitation programme in one South-East Asian country, it was discovered during the planning analysis phase that many coconut farmers were not adopting the recommended new and improved MAWA seed (a hybrid of yellow Malaysian Dwarf and West African Tall species of coconuts). Their unresponsiveness was initially the main concern of the extension activity. It was found that a large number of coconut farmers were already well-informed about this type of seed and they were favourably disposed towards using it. However, many coconut farmers reported that they felt there was no need to increase their coconut yields because the price of copra was very low. After further study and analysis, it was found that the low copra price was attributed to the poor copra quality, which was due to the ineffective method of drying the coconut. Thus, one of the main emphases of the campaign on this coconut replanting and rehabilitation programme should have been on informing or educating coconut farmers on new or improved coconut-drying techniques and methods rather than on promoting the improved MAWA seeds. If such appropriate (i.e. simple and inexpensive) technology for effective coconut drying is not available, then it is also the task of extension planners to sensitize and encourage researchers, technicians, decision-makers or administrators involved in coconut production to resolve that problem. Thus the target audience of extension campaign activities do not always have to be the "farmers" or "rural" population; when necessary, they can also include the "big bosses".

In one South Asian country, a programme of voluntary female sterilization (ligation) was launched. Before a systematic baseline/KAP survey was conducted almost all of the programme's decision-makers, planners and managers thought (or assumed) that the main problem of the programme was that most couples, especially the husbands, were afraid that their sex-life might be affected after sterilization. However, KAP survey results revealed that the programme organizers' perception of the problem was incorrect. Almost two-thirds of the respondents reported that the main reason for not having sterilization was due to their "fear of operation" (the first-ranking reason) and only 26.5 percent cited "sex-life might be affected after the operation" (ranking fourth). Had the KAP survey not been conducted, the programme thrust in terms of campaign strategy and message positioning would have been mainly on counter-attacking "the negative effects on sex-life" rather than the "fear of operation" issue.

The foregoing two simple illustrations, and also the examples provided earlier on the planning process of the rat control campaign in Malaysia, indicate the need to analyse and identify specific and main issue(s) or problem(s) which an extension programme has to resolve. Once such problems have been identified, specific campaign strategies need to be developed to solve each of the problems encountered by the programme. Figure 2-5 provides simplified guidelines on how to determine the general campaign strategy direction and/or priority on the basis of KAP survey findings. These general strategies need to be made more specific in Phases 4, 5 and 6 of the campaign planning process, which is discussed in greater detail later. It should be mentioned that the general and simplified guidelines should not be applied rigidly in all situations. Rather, it should be utilized as a tool to conceptualize and systematize a campaign planning and strategy development process.

FIGURE 2 - 5

General and Simplified Guidelines on Utilizing Results of Knowledge, Attitude and Practice (KAP) Survey for Planning and Development of Extension Campaign Strategies.

SITUATION	IF — Level of Audience's: K Knowledge	A Attitude	P Practice	THEN — GENERAL/BROAD EXTENSION STRATEGY PRIORITY: Main Approach	Main Purpose	Information Positioning	Communication Channel Emphasis for Multi-Media Mix Usage: Mass	Inter-personal	Group	Supplies/inputs & services availability
1	Low	Low	Low	INFORMATIONAL	CREATE AWARENESS AND INCREASE KNOWLEDGE	"WHAT"	High	Low	Low	DESIRABLE
2	Medium	Low	Low	INFORMATIONAL	INCREASE KNOWLEDGE; EXPLAIN NEEDS FOR AND BENEFITS OF ADOPTING THE IDEAS	"WHAT" AND "WHY"	High	Low	Low	DESIRABLE
3	High	Low	Low	MOTIVATIONAL	EXPLAIN BENEFITS, REWARDS AND USEFULNESS OF SUGGESTED IDEAS, INFORM CONSEQUENCES OF NON-ADOPTION	"WHY"	High	Low	Moderate	IMPORTANT
4	Medium	Medium	Low	MOTIVATIONAL AND EDUCATIONAL	AS ABOVE	"WHY"	High	Moderate	Moderate	IMPORTANT
5	Medium	Medium	Medium	MOTIVATIONAL AND ACTION	CHANGE UNFAVOURABLE PERCEPTIONS NEUTRALIZE NEGATIVE ATTITUDES AND BELIEFS COUNTER-ATTACK MISCONCEPTIONS	"WHAT" AND "WHY"	Moderate	Moderate	High	ESSENTIAL
6	High	Medium	Low	MOTIVATIONAL AND EDUCATIONAL	INFORM CONSEQUENCES OF NON-ADOPTION CORRECT MISUNDERSTANDING	"WHY" AND "HOW"	Low	High	Moderate	VERY ESSENTIAL
7	High	Medium	Medium	MOTIVATIONAL AND EDUCATIONAL	AS ABOVE	"WHY" AND "HOW"	Low	High	Moderate	ESSENTIAL
8	High	High	Low	EDUCATIONAL AND ACTION	PROVIDE LOGISTICAL INFORMATION FOR ACTION DEMONSTRATE & TEACH PROPER USE OF RECOMMENDED TECHNOLOGY	"HOW"	Low	High	High	VERY ESSENTIAL
9	High	High	Medium	EDUCATIONAL AND ACTION	DEMONSTRATE & TEACH PROPER USE OF RECOMMENDED TECHNOLOGY	"HOW"	Low	High	High	ESSENTIAL

Source: Adhikarya, (1985).

Note: The classification of KAP levels (high, medium, low) depends on the type of innovation/idea promoted, size of population, cost/effort required, etc. It's a rather arbitrary judgement as to what criteria to use to classify these levels, but normally low level is 0-30%, medium level is 31-60%, and high level is above 60% for knowledge or attitude, and 0-20% (low), 21-40% (medium), and 40% (high) for practice.

Phase 4:   *Audience analysis and segmentation*

One of the most important elements in a campaign is the target audience or beneficiaries : who they are, where they are located, why they are chosen as target beneficiaries and what information contents or messages should be communicated to them. Analysis of a target audience is an integral part of designing and planning a campaign strategy. In such an analysis, certain types of information or data are needed; for instance, the size and location of the target audience, their socio-economic profile (including age group, income, occupation, education, among other data) and their socio-cultural profile (including religion, language, family life patterns, traditional belief systems, norms, values, information sources, communication and interaction practices, among others). Characteristics, interests and information needs of the target audience might be different, so audience segmentation into several different target groupings is usually necessary. For each target group, a specific campaign strategy may be required. In this phase, it is also important to prioritize the target beneficiaries as to which group(s) should be reached first or be given the most intensive campaign treatment.

Tables 2-6 and 2-7 are examples of the results of a KAP survey on weed management conducted in 1988 among farmers in Malaysia's Muda Agricultural Development Authority (MADA). Based on these data, useful strategies for campaign planning, including priority target group, location, type/nature of information needed, etc. can be developed. Table 2-6 revealed that most farmers surveyed did not use the correct amount of Arrosolo, while among those farmers using Rumputox only about half of them applied the recommended amount. In the use of Ronstar, about 81 percent of farmers who were located in Districts III and IV of the MADA scheme used the correct amount, while the majority (84.4 percent) of those in Districts I and II did not use the recommended amount. If one follows a strategic planning principle, one possible option is to concentrate the SEC activities on providing information, motivation and education to farmers in all districts on the correct amount of Arrosolo use, and the correct amount of Ronstar use only to farmers in Districts I and II.

TABLE 2-6

Amount of herbicide applied by farmers in the Muda irrigation scheme, Malaysia (1988)

Type of herbicide	Amount of herbicide used for each relong		Percentage of farmers in	
			District I and II	District III and IV
RUMPUTOX	< 250	gr	10.1 %	11.1 %
	251 - 500	gr	31.0 %	23.4 %
	501 - 750	gr	1.0 %	5.3 %
	751 - 1000	**gr***	**46.1 %**	**49.1 %**
	> 1000	gr	12.2 %	11.1%
ARROSOLO	<1000	gr	100.0 %	72.7 %
	1001 - 1500	ml		18.2 %
	1501 - 2000	**ml***		**9.1 %**
	> 2000	ml		
ORDRAM	< 5	kg	60.8 %	42.1 %
	5 - 7.5	kg	4.3 %	17.5 %
	7.6 - 10	**kg***	**21.7 %**	**24.6 %**
	> 10	kg	13.0 %	15.8 %
RONSTAR **	< 1000	ml	5.4 %	3.0 %
	1001 - 1500	ml	10.8 %	5.5 %
	1501 - 2000	**ml***	**15.6 %**	**80.7 %**
	> 2000	ml	68.2 %	10.8 %

* the correct/recommended amount   ** For Ronstar, we use simulated data for illustration only

Source: R. Mohamed and Y.L. Khor, "Survey Report of Farmers' Knowledge, Attitude, and Practice (KAP) on Weed Management in the Muda Agricultural Development Authority (MADA)", Malaysia (March 1988).

TABLE 2-7

Timing of herbicide applications by farmers in the Muda irrigation scheme, Malaysia (1988)

Type of herbicide	Location	Percentage of farmers who applied herbicide				
		0-4 days	5-7 days	8-10 days	11-15 days	>15 days
				after sowing		
RUMPUTOX	District I & II	1.7%	2.8%	3.9%	32.0%	**59.6%***
	District III & IV	0.6%	2.3%	2.3%	28.7%	**66.1%***
ARROSOLO	District I & II	8.3%	8.3%		33.3%	**50%***
	District III & IV			40.0%	30.0%	**30%***
ORDRAM	District I & II	4.2%	12.5%	12.5%	**29.2%***	41.7%
	District III & IV	1.6%		8.2%	**37.7%***	52.5%

* the correct/recommended timing

Source: R. Mohamed and Y.L. Khor, "Survey Report of Farmers' Knowledge, Attitude, and Practice (KAP) on Weed Management in the Muda Agricultural Development Authority (MADA)", Malaysia (March 1988).

Attitudes of Rice Farmers in Penang Towards Rat Control

Statements regarding rat habits and control measures (selected statements only)	Rice farmers' attitude				
	strongly agreed	agreed	neutral	disagreed	strongly disagreed
Rats are intelligent thus rat control will not succeed	8%	44%	17%	30%	1%
A group effort to eliminate rats is sadistic	3%	27%	8%	55%	7%
Rats will take revenge on behalf of their dead friends by causing worse damages	12%	42%	27%	**17%**	**2%**
Rat control is farmers' responsability and not that of the Government/Dept. of Agriculture	10%	81%	2%	7%	0%
Simultaneous planting is not an important factor in rat control	6%	14%	1%	43%	36%

Source: A. Hamzah and J. Hassan, "Rice Farmers' Knowledge, Attitude, and Practice of Rat Control: A study conducted in Penang, Malaysia", Serdang: Agricultural University of Malaysia, (August 1985).

Table 2-7 also suggests that higher priority should be given to providing farmers with more and better information, motivation and education regarding the correct timing of Ordram use (i.e., 11-15 days after sowing) as more than two-thirds of farmers surveyed did not follow the recommended timing. In terms of timing for Arrosolo use, the data also indicated that more attention should be given to farmers in Districts III and IV of whom only 30 percent applied the herbicide at the correct or recommended time.

KAP survey results regarding farmers' attitudes on rat control in Penang also serve as good illustrations on how non-technological factors can hinder the adoption of recommended technologies. As shown in Table 2-8, superstitious belief was one of the dominant reasons for not controlling rats as only 19 percent of all farmers surveyed disagreed that "rats will take revenge on behalf of their dead friends by causing worse damages". No matter how good the recommended technology for rat control is and how well informed these farmers are on the methods of rat control, if the majority of them had such a superstition, the success of a rat control campaign is unlikely unless attitudinal change efforts to counter-attack or neutralize such a misperception are undertaken effectively as part of the campaign.

More detailed real-life examples of audience analysis and segmentation are provided below. Although these examples are not agriculture-related, they represent clear illustrations of important SEC processes and principles

TABLE 2-9

Percentage of Ever-Married Women and Men Under 50 Years of Age Knowing about and Having Undergone Vasectomy and Tubectomy, by Geographical Division, 1981

Division	Tubectomy		Vasectomy	
	Knowledge	Practice	Knowledge	Practice
Rajshahi	94.3	2.8	80.5	0.9
Khulna	96.9	4.6	80.3	1.0
Dhaka	95.7	5.5	68.8	0.9
Chittagong	82.5	1.7	58.6	0.4

Source: Compiled from Bangladesh Contrapceptive Prevalence Survey, (1981)

TABLE 2-10

Percentage of Current Users Among Ever-Married Women Under 50 Years of Age Practising Specific Family Planning Methods, by Educational Level

Education	Tubectomy	Vasectomy	Pill	Condom
None	22.0	11.6	26.3	6.6
Some primary	21.5	2.6	35.6	13.8
Completed primary	14.8	1.1	32.8	24.6
Higher	3.9	0.0	35.0	31.3

Source: Compiled from findings in Bangladesh Contraceptive Prevalence Survey, (1981)

TABLE 2-11

Average Number of Children Ever Born and Number of Living Children Among Ever-Married Women and Men Under 50 Years of Age, by Age Group

Age group	Number of	
	Ever-born	Living children
<15	0.1	0.1
15-19	0.7	0.6
20-24	2.1	1.7
25-29	3.7	2.9
30-34	5.4	4.2
35-39	6.4	4.9
40-44	7.3	5.3
45-49	7.6	5.3

Source: Compiled from Bangladesh Contrapceptive Prevalence Survey, (1981)

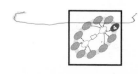

which can be applied to many agricultural extension activities. The following examples are from a population education campaign to increase tubal ligation (female sterilization by tubectomy) acceptors in Bangladesh (Adhikarya, 1983). In analysing and segmenting the target audience of the ligation campaign, findings from baseline/KAP surveys were utilized.

Results of the 1981 Bangladesh Contraceptive Prevalence Survey were used in analysing the target audience for the ligation campaign. The data showed that urban women (5.7 percent) had a higher preference for ligation over rural women (3.5 percent). Also, ligation acceptors from urban areas in 1981 consisted of 36.8 percent of the total number of ligation acceptors, compared with 63.2 percent from rural areas. Considering the national urban-rural proportion, there seemed to be an urban concentration in the ligation programme in Bangladesh. The programme organizers thus suggested greater emphasis and more intensive efforts on reaching and motivating rural women.

Table 2-9 shows the relatively poor performance of the ligation programme in Chittagong Division, both in terms of knowledge level (82.5 percent) and actual practice (1.7 percent) as compared to Khulna and Dhaka Divisions. In Rajshahi Division, while the knowledge level was quite high (94.3 percent), the actual practice level was relatively low (2.8 percent) as compared with Khulna (4.6 percent) and Dhaka (5.5 percent) Divisions. Those findings indicated that additional efforts should seek to increase the audience's information exposure in order to raise their knowledge or awareness levels regarding tubal ligation and to launch specific motivation and persuasion activities to increase the adoption level of tubal ligation in Chittagong and Rajshahi Divisions.

Another interesting finding is reflected in Table 2-10: with both tubectomy and vasectomy (male sterilization) methods, unlike other contraceptive methods, the percentage level of acceptors decreases as the level of education increases. For pill and condom acceptors, on the contrary, the percentage level of acceptors increases with the increase in educational level. While tubectomy seems to be preferred more by urban women, the educational level of those women seems to be low.

The Bangladesh Contraceptive Prevalence Survey of 1981 reported that most of ligation acceptors were in two age groups: 30 to 34 years (7.1 percent) and 35 to 39 years (6.6 percent). However, the data also indicate that among those two age groups the number of living children, as well as the number of ever-born children, was already high (see Table 2-11). To achieve a better demographic impact, it was necessary for the ligation campaign to concentrate on the 20 - 24 and 25 - 29 age groups whose average number of living children was lower (1.7 and 2.9 children, respectively) compared with the 30 - 34 and 35 - 39 age groups (4.2 and 4.9 children, respectively).

Since a large number of ligation acceptors (43 percent) have five or more children, it seems that most couples want a certain guarantee that they have at least four living children. The data also indicate that a considerable number of ligation acceptors had their last living child between two and five years earlier. It thus appeared that unless couples were convinced that their infants or young children below the age of five had a good chance of surviving, they would be less likely to accept ligation, especially among women who were in the younger age group (e.g., 20 - 24 or 25 - 29 years), or at the lower parity group (i. e., having two or three living children).

The 1983 KAP survey conducted by the Bangladesh Association for Voluntary Sterilization (BAVS) reported that 78 percent of BAVS ligation acceptors had two or three sons, compared with only 56 percent who had two or three daughters. It seems that most women were more likely to accept ligation when they already had at least two sons. The data suggest that a "son preference" existed among ligation acceptors in Bangladesh.

On the basis of the above audience analysis, some strategy implications for the ligation campaign were identified :

   a   Reach rural women with relatively low education (e.g., primary level or lower), especially those located in Chittagong and Rajshahi.

   b   Reach urban women with relatively high education (e. g., completed primary level or higher).

   c   The primary target audience should be in the 20 - 29 age group and having not more than three living children.

   d   Inform the target beneficiaries that, due to better health facilities and conditions, infant and/or child mortality incidents have declined and chances of infant or child survival below five years of age have improved significantly.

   e   Discount/counter-attack the belief that sons are better than daughters.

Another way of analysing and segmenting target groups is by clustering target beneficiaries' perceived problems according to the reasons for not having adopted the recommended campaign messages or suggestions. For instance, the problems can be evaluated in terms of relative degree of difficulty in their solution, thus placing each problem into a scale which has a continuum from "less difficult" to "more difficult". Figure 2-12 illustrates how three different target groups - the Motivated Group, the Sceptical Group and the Resistant Group - in the ligation campaign in Bangladesh were segmented on the basis of their perceived problems of non-adoption of the ligation method (Adhikarya, 1983).

The three different target groups were identified according to three different clusters of non-adoption problems. Message design, development and delivery for these groups were thus facilitated and planned with greater precision based on their specific information needs. For instance, information or messages required for Target Group I (Motivated Group) was more of a logistical nature, i.e., location and opening hours of clinics offering tubectomy services, duration of the operation and recuperation, whether the operation is free and/or compensation will be given for wage loss and travel expenses, etc. For Target

FIGURE 2-12

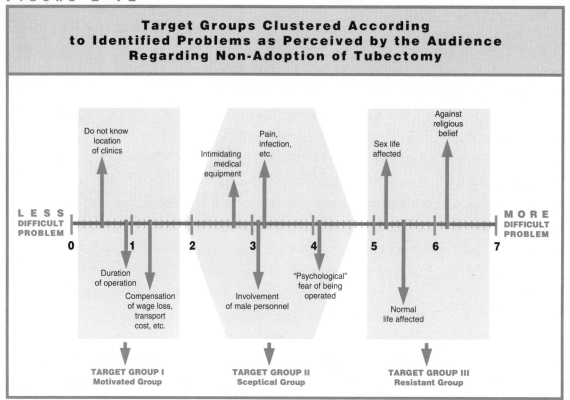

Target Groups Clustered According to Identified Problems as Perceived by the Audience Regarding Non-Adoption of Tubectomy

Source: Adhikarya (1983)

Group II (Sceptical Group) the information had to neutralize or correct their misconceptions regarding ligation, such as the fear of operation, including the side- or after-effects, infection, haemorrhage, pain, fever, intimidating medical equipment, whether a male or female nurse would perform the operation, and others. The information for Target Group III (Resistant Group) had to refute and counterattack certain mistaken beliefs and opinions regarding ligation, vis-a-vis that physical strength or fitness and sex-life would be negatively affected after the operation, that ligation was sinful or prohibited by religion (i.e., Islam), among others.

If a strategic decision had to be made contingent on limited resources or time, it would seem quite reasonable for the campaign to concentrate on the Motivated Group since the chances of success would appear to be greater with that target group than with the two others. A campaign success with the Motivated Group could thereafter be used by the campaign organizers to persuade policy-makers or funding agencies to provide more resources for dealing with the next (and more difficult) target group, the Sceptical Group, and so on. Audience analysis and segmentation can thus facilitate planning for a phased campaign approach in order to permit less difficult problems to be solved first and the more difficult ones later with increased resources and experience.

An analysis of the target audience characteristics such as access to communication channels, information-seeking habits, preferred information sources, patterns of media usage, communication network interactions and group communication behaviour would also be very important and useful in selecting a cost-effective combination of multi-media channels and in planning the most appropriate use of such a media mix to support the campaign activities. That process of multi-media selection and planning is discussed in greater detail in the next phase.

Phase 5: *Multi-media selection*

Many campaign experiences and empirical research studies on communication media effects have shown that the use of multiple media channels which include a combination of mass, group and interpersonal communication, if appropriately selected and utilized, is usually more cost-effective than the use of a single communication medium. An important aspect in employing a multi-media approach is the proper selection of available channels in order to avoid redundant or superfluous media usage and to optimize the level of multi-media support required. Thus, a multi-media approach does not mean that all available communication channels should be utilized.

General guidelines for selection of multi-media mix should be based on specific campaign objectives and strategy, KAP levels, etc., as suggested in Figure 2-5. For instance, the degree of emphasis in utilizing mass, interpersonal or group communication channels depends on target beneficiaries' KAP levels and campaign strategy priorities. In developing an appropriate multi-media mix, results of audience analysis should be considered, especially on information-seeking habits, preferred information sources, media access or ownership, media consumption or usage patterns, communication network interactions and group communication behaviour.

The rationale behind the use of a multi-media approach is that a coherent, coordinated, and reinforcing system of communication should be able to address specific but varied information problems and needs of target beneficiaries. In addition, since there is no information medium which is effective for all communication purposes or for all types of target beneficiaries, a multi-media approach is considered as a viable alternative. Another reason for employing a multi-media approach is the need to make the extension system more efficient given the various information, educational and communication objectives of a campaign. For example, as shown in Figure 2-5, if the main purpose of a campaign is create awareness and increase the knowledge level of target beneficiaries, the most efficient and effective way is to utilize mass communication channels and not interpersonal or group communication approaches.

One of the most important factors in employing a multi-media approach effectively is strategic planning on which medium or combination of media should be used for what specific purpose by whom, in order to deliver which specific messages or information to whom. To illustrate more clearly the process of multi-media strategy planning, two examples are recounted here from the Bangladesh Ligation Campaign (Adhikarya, 1983) which were aimed at Target Group I (Motivated Group; Figure 2-13) and Target Group II (Sceptical Group; Figure 2-14).

For Target Group I, as shown in Figure 2-13, there were two main target sub-groups (a) urban women and (b) rural housewives. The multi-media strategy plan suggested that the two sub-groups be reached by two types of intermediaries: urban men, especially industrial (or semi-skilled) labourers and (b) rural men. The intermediaries would be exposed to campaign information or messages through various channels in strategic locations or places where they frequently were to be found, such as in factories, industrial plants, government offices, clinics, local dispensaries and cinemas. Having been exposed to campaign messages, the intermediaries would be expected to communicate with and influence their wives regarding ligation. In addition to the radio spots aimed at both target groups, mini-posters would be pasted on matchboxes which were used in most urban and rural households. The systems activity chart (Figure 2-13) of the multi-media strategy plan describes the selected multi-media inputs, the management or delivery system, and the specific target groups for whom the messages and media were intended.

For Target Group II (see Fig. 2-14), a similar approach as that of Target Group I was proposed: the use of intermediaries such as rural men and female BAVS field agents to reach rural women, and primary school children and BAVS clinic counsellors to reach rural men. Mini-posters would be pasted on coconut-oil bottles and distributed by commercial manufacturers to all parts of Bangladesh, including the most remote rural areas. Since coconut oil is used by a majority of low-income population for their hair, "piggy-backing" on an essential and inexpensive commercial product which has well-established distribution outlets and procedures would likely be more cost-effective than attempting to distribute leaflets or posters in a conventional way.

Since there are many different factors which need to be considered for an appropriate media-media mix, standardized guidelines for a multi-media selection process might prove to be counter-productive. A general approach which might assist in the selection of multi-media channels includes the following :

a	Use a medium for a single or specific purpose rather than for several different goals.
b	Select a medium which has a unique characteristic or particular advantage which is useful to accomplish a specific purpose.
c	Select a medium which the target audience is already familiar with and has access to.
d	Use a medium which can easily accommodate "localized" messages, if necessary.
e	Select a medium for which operational support is locally available, and the materials can be developed and produced locally.
f	Use a combination of media which can complement and reinforce each other but have different main functional strengths or emphases.

While the process of multi-media selection is important, the effectiveness of the media chosen also depends on how appropriately the contents to be conveyed by the media are designed, developed, and packaged. The next section discusses the process of message design, development, pretesting and multi-media materials production.

FIGURE 2-13

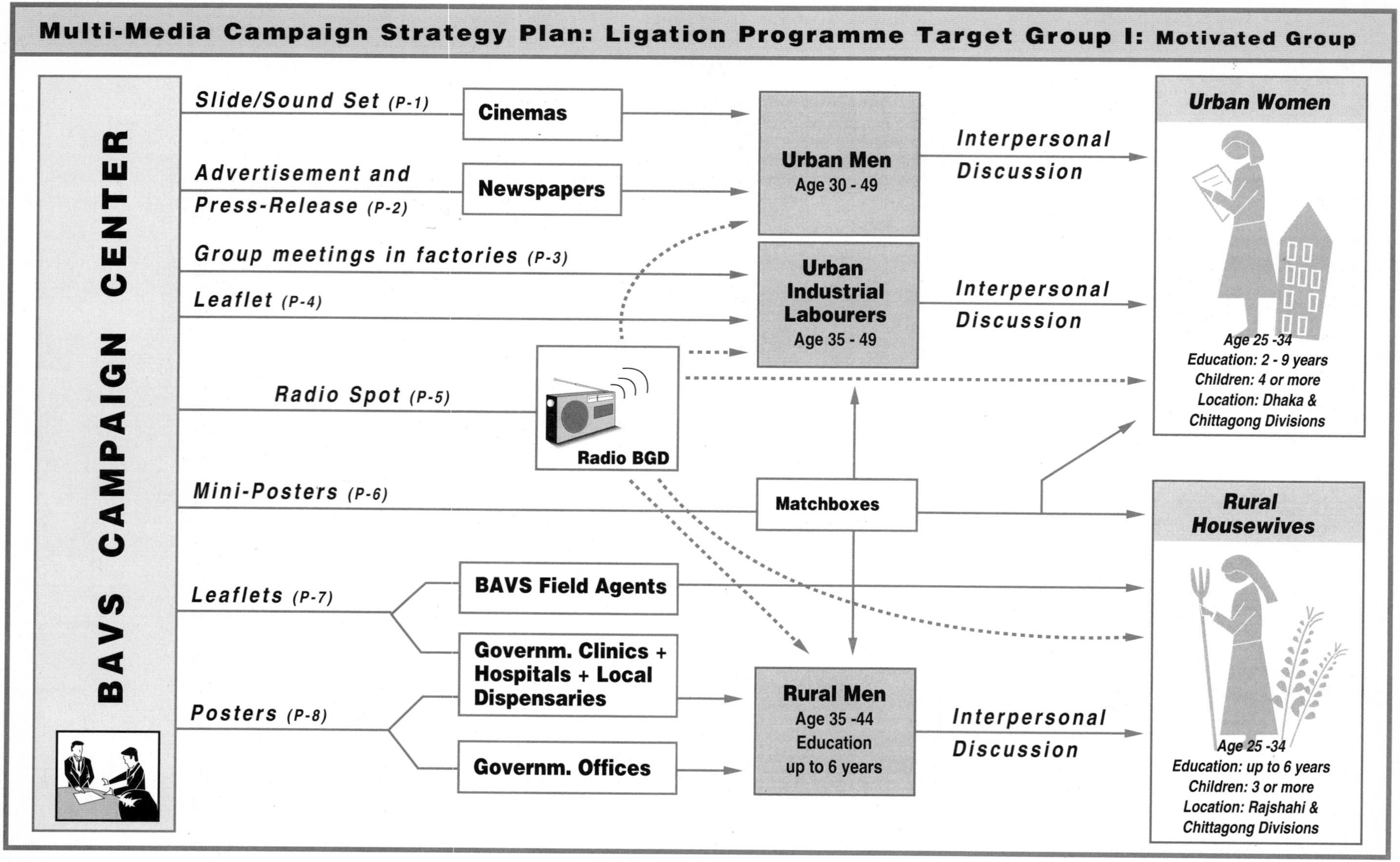

Multi-Media Campaign Strategy Plan: Ligation Programme Target Group I: Motivated Group

Urban Women

Age 25 -34
Education: 2 - 9 years
Children: 4 or more
Location: Dhaka &
Chittagong Divisions

Rural Housewives

Age 25 -34
Education: up to 6 years
Children: 3 or more
Location: Rajshahi &
Chittagong Divisions

Interpersonal Discussion

Interpersonal Discussion

Interpersonal Discussion

Urban Men Age 30 - 49

Urban Industrial Labourers Age 35 -49

Matchboxes

Rural Men Age 35 -44 Education up to 6 years

Cinemas

Newspapers

Radio BGD

BAVS Field Agents

Governm. Clinics + Hospitals + Local Dispensaries

Governm. Offices

Slide/Sound Set (P-1)

Advertisement and Press-Release (P-2)

Group meetings in factories (P-3)

Leaflet (P-4)

Radio Spot (P-5)

Mini-Posters (P-6)

Leaflets (P-7)

Posters (P-8)

BAVS CAMPAIGN CENTER

Source: Adhikarya (1983)

Selected Media for Target Group I:
Motivated Group (see Figure 2-13)

Cinema Slide-Shows (Product 1 or P-1)**:** Three or four sequential slides to be screened at cinemas in urban areas, and if possible accompanied by a short jingle or commentary and song, are proposed for reaching urban men.

Newspapers (Product 2)**:** Special advertisements and feature articles (part of publicity/press release activities) are proposed for reaching urban men.

Group meetings (Product 3)**:** Organized group meetings for urban industrial (semi-skilled) male labourers should be conducted in their place of work with the cooperation of their employers (e.g. jute mills, garment factories, pharmaceutical plants). Such an approach would facilitate the task of gathering the specific target audience and seems to be a practical and efficient way of establishing a two-way communication channel.

Leaflets: Two types of leaflets should be developed for three types of audience groups.
The first leaflet (Product 4) is to support the group meeting activities (summary of the important points/messages) for urban industrial male labourers who can take it home and keep it as reference material.
The second leaflet (Product 7) is for rural women (to be distribuited through BAVS field agents) and rural men (to be distributed through government clinics, hospitals and local dispensaries). The contents of this second leaflet should reinforce the radio messages so that the target group could remember or refer back to the important radio broadcast which otherwise could easily be forgotten.

Radio (Product 5)**:** Short but attractive radio spots (30 or 60 seconds, four times daily) should be aimed specifically at both urban women and rural housewives who otherwise are difficult to reach directly. In addition, another radio spot should aim at both rural and urban men, to reinforce the messages contained in other printed or visual media.

Mini-poster on matchboxes (Product 6)**:** Short messages, slogans, logo, etc., can be disseminated most effectively and efficiently through pasting or labelling a mini-poster on commercial matchboxes. Since these matchboxes have already their own commercial outlets, this "piggy-backing" approach (by paying the labelling charges to the matchbox manufacturers) could avoid many management and distribution problems.

Posters (Product 8)**:** Visually attractive posters should be developed for rural educated men containing relevant localized information or messages. Posters could be printed with nation-wide campaign messages, but leaving a blank space on which local BAVS clinics could write (with felt pens) their relevant localized information or messages. Such posters should be placed at (a) government clinics, hospitals and local dispensaries and (b) outside government offices.

Source: Adhikarya (1983)

FIGURE 2-14

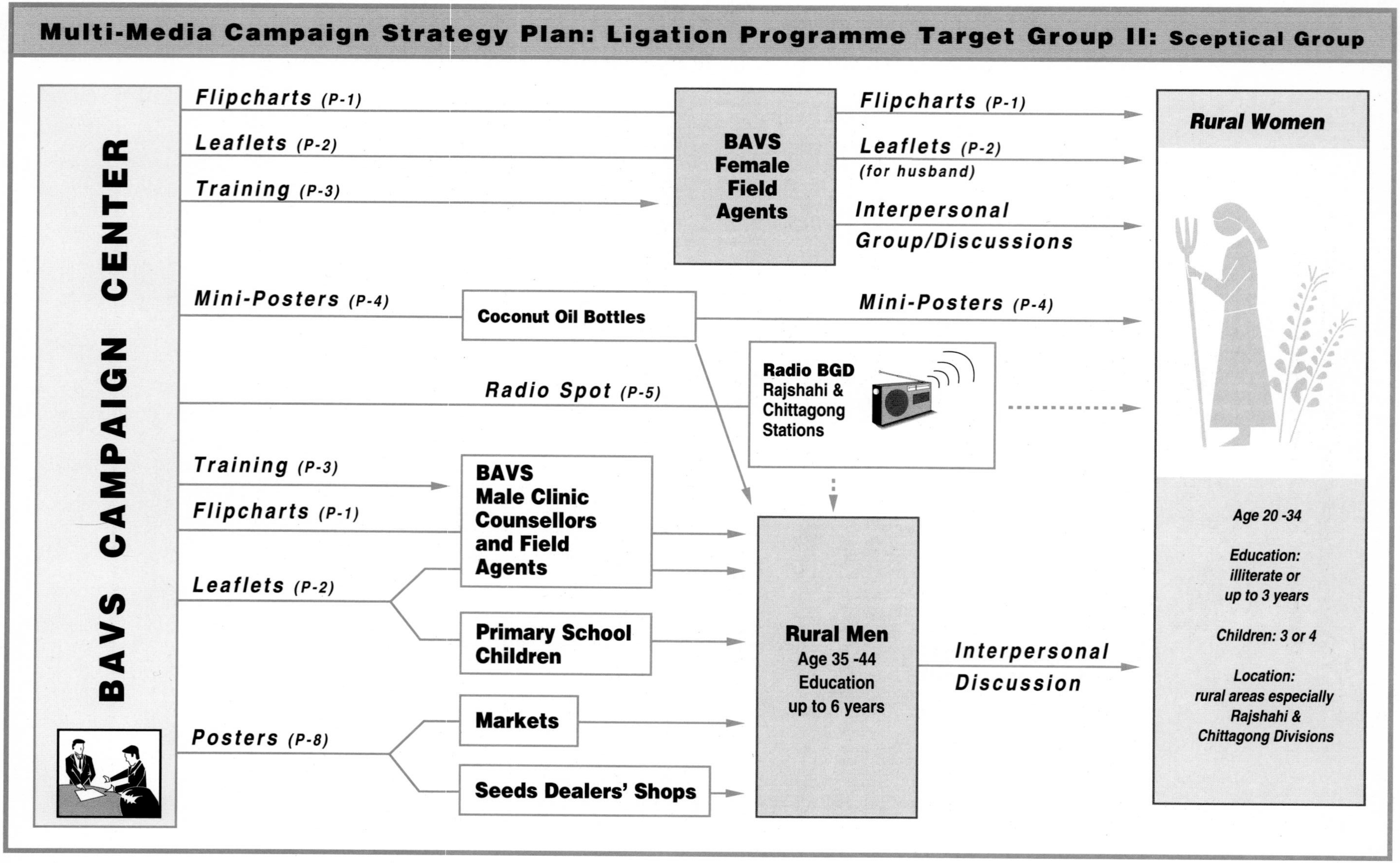

Multi-Media Campaign Strategy Plan: Ligation Programme Target Group II: Sceptical Group

Rural Women

Age 20 -34

Education: illiterate or up to 3 years

Children: 3 or 4

Location: rural areas especially Rajshahi & Chittagong Divisions

Flipcharts (P-1)

Leaflets (P-2) (for husband)

Interpersonal Group/Discussions

Mini-Posters (P-4)

BAVS Female Field Agents

Coconut Oil Bottles

Radio Spot (P-5)

Radio BGD Rajshahi & Chittagong Stations

Interpersonal Discussion

Rural Men Age 35 -44 Education up to 6 years

BAVS Male Clinic Counsellors and Field Agents

Primary School Children

Markets

Seeds Dealers' Shops

Flipcharts (P-1)

Leaflets (P-2)

Training (P-3)

Mini-Posters (P-4)

Training (P-3)

Flipcharts (P-1)

Leaflets (P-2)

Posters (P-8)

BAVS CAMPAIGN CENTER

Source: Adhikarya (1983)

Selected Media for Target Group II:
Sceptical Group (see Figure 2-14)

Flipcharts (Product 1 or P-1)**:** Flipcharts containing specific informational and motivational messages should be developed for use by BAVS counsellors and field agents during organized group or individual meetings with rural women (primary target audience group) and rural men (intermediaries).

Leaflets (Product 2)**:** Information contained in this leaflet will be distributed through three channels:
(1) BAVS women field agents (for rural women)
(2) BAVS male consellors and male field agents (for rural men)
(3) Rural primary school children (who will receive the leaflet in a sealed envelope addressed to their fathers).

Training (Product 3)**:** Special training for BAVS male and female field agents and counsellors should be conducted and relevant training materials developed regarding the specific campaign activities (messages, use of flipcharts, distribution of leaflets, etc.)

Mini-poster on coconut-oil bottles (Product 4)**:** The idea is similar to that of using mini-posters on matchboxes (as for the Target Group I) but, as more space is needed for the Target Group II messages, the mini-poster should be of about 2"x3" in size, and to be pasted on the empty space of coconut-oil bottles. This coconut oil is used by the majority of the population in Bangladesh, both rural and urban men and women, to put on their hair, and is inexpensive.

Radio Spots (Product 5)**:** Entertaining yet informative and motivational radio spots (60 seconds) should be aimed at both rural men and women. Such radio broadcast by Radio Bangladesh (BGD) should reinforce the messages of the leaflets, flipcharts and posters.

Posters (Product 6)**:** Short messages summarizing the salient points of information needed for Target Group II should be included in the posters aimed at rural men. Such posters should be placed at the markets and seed dealers' shops which most rural men visit frequently.

Source: Adhikarya (1983)

Phase 6:   *Message design, development, pretesting and materials production*

The effectiveness of a campaign, to a large extent, depends on the relevance, validity and practicality of the information or messages communicated to the target beneficiaries. Sometimes, even though the main content of the information is useful and technically sound, the campaign information may not be well received or understood by the target beneficiaries if it is not presented properly to them. Campaign messages must be able to attract the attention of the target audience, be easily and clearly understood, and be accurately perceived by them.

In general, it is often assumed that the degree of message effectiveness is a function of the amount of reward that the message offers and the level of efforts required by the audience to interpret, perceive and understand the message. In message design and development processes, given the formula:

$$\textbf{M.e.} \text{ (Message effectiveness)} = \frac{\textbf{R} \text{ (reward)}}{\textbf{E} \text{ (efforts)}}$$

then the option that is often more within the control of extension campaign planners is to decrease the level of efforts required by the target audience in interpreting, perceiving and understanding the campaign messages.

Another factor which may influence the effectiveness of a campaign message delivery is the competing messages that try to reach the same target audience. Rural population in many developing countries are "target beneficiaries" of numerous social, economic and political development programmes which tend to overburden them with various types of information. To make matters worse, the messages often conflict with one another, thus further confusing target beneficiaries. For instance, a nutrition campaign message may suggest "establish water/fish ponds" while a health campaign message may recommend "get rid of stagnant water" (for disease control).

Given the "information overload" situation, which has become increasingly common among rural population in many developing countries, a campaign message needs to be strategically "positioned" in order to "stand out" in the crowd. Otherwise a message may go unnoticed and has no impact even though it is technically useful and relevant for the target audience.

Positioning a campaign message strategically and effectively into target beneficiaries' minds requires a well-planned, creative and innovative approach of presenting the message. First of all, the relevance and validity of the message must be ensured. On the basis of a KAP survey results, the broad or general campaign approach (i.e. informational, motivational or action) should be identified, as outlined in Figure 2-5. Then a campaign message focus or theme must be identified according to the specific campaign issue or objective. In a campaign, more than one theme may be necessary, depending on the campaign objectives and information needs.

Once a theme has been identified, messages need to be developed and treated further for effective "packaging", utilizing various social and psychological appeals, in order to make the theme attractive and persuasive. Depending on the campaign objectives and strategy, the following examples of the more commonly used social and psychological appeals can be applied in packaging campaign messages:

→ fear-arousal appeals   → authoritative appeals
→ incentive/rewards appeals → emotional appeals
→ testimonial appeals   → civic-duty appeals
→ community or group/   → morale-boosting appeals
  peer pressure appeals   → common-man appeals
→ role-model appeals   → guilt-feeling appeals
→ etc.

Messages which have been packaged with appropriate appeals can be given different treatments according to the needs, objectives, and strategies of a campaign. For instance, a message presentation can be given one or more of the following treatments:

→ serious/formal
→ humorous
→ popular/informal
→ one-sided (i.e. pros or cons only)
→ two-sided (i.e. pros and cons)
→ repetitive

→ aggressive/confrontational
→ direct
→ indirect
→ conclusion-drawing
→ fact-giving
→ etc.

The packaging of messages using the above types of presentation appeals and treatments should also take advantage of, or capitalized on, the particular strength or positive attributes of the intended delivery medium. For instance, the advantages of radio differ from those of a leaflet or a filmstrip, and so on. Each medium has its own features which can be utilized to different effects through appropriate message design with the appropriate appeals and treatments. In general, message appeals using fear-arousal or emotional approach, for example, may be best expressed by a visual medium, whereas fact-giving or conclusion-drawing types of message presentation treatment are more likely to be effective if presented through a printed medium. Therefore, in designing and developing messages, the intended medium to be utilized to deliver the message must be also be taken into consideration.

In order to ensure that target beneficiaries correctly interpret, perceive and understand the meaning of a message, a pretesting exercise would be very useful and should be undertaken. Such an exercise is needed especially if visual materials are intended for an illiterate or lowly-educated audience (i.e. less than three years of formal schooling) who might also be illiterate visually because symbolic representation must be learned and experienced in order to be understood. A message which has been developed and packaged on a prototype basis should be pretested before final production with a small sample of the actual target beneficiaries. A message pretesting exercise can be done in a very short time with limited cost and simple methodology, and may result in a significant improvement in the effectiveness of the message as well as a considerable saving of resources (time, effort and funds). Such formative evaluation of campaign materials (i.e. messages and the delivery medium) prior to the actual implementation of a campaign activity, as suggested in Figure 2-1, should thus be considered as a built-in or integral part in a campaign strategy planning process.

Phase 7:   *Management planning*

A good campaign strategy plan does not automatically result in an effective campaign implementation. Unless a management plan is drawn up to specify how the strategy should be put into operation, a campaign is likely to be unsuccessful because the required logistical arrangements and support materials may not have been made or completed on time or according to the specifications of the strategy plan.

There are at least two major elements with which a management plan should be concerned. The first element deals with the assignment of responsibility regarding the specific tasks to be undertaken to conduct the field implementation of the strategic campaign plan. Detailed guidelines and/or activities in mobilizing the necessary resources (staff, logistical support, funds, etc.) to support the campaign implementation plan should be specified in such a plan. Another element of a management plan deals with the organization and coordination of planned multi-media activities as well as the proposed system of campaign materials distribution and utilization. Guidelines on the specific recipients and locations, as well as the utilization procedures of such multi-media support materials should be provided in the management plan. These guidelines and procedures are not only needed to facilitate campaign field implementation, but also needed for management monitoring purposes.

As discussed in the beginning of this chapter, one of the main objectives of management planning is to provide campaign organizers with a systematic and comprehensive procedure for mobilizing available resources (personnel, finances and time) effectively and efficiently according to the campaign strategy plan. Since campaign implementation depends to a large extent on persons who are mobilized to carry out the different aspects of the campaign activities, a plan for managing training activities according to the specific campaign training needs is also required. The implementation of a multi-media extension campaign can only follow the planned strategy if: (a) the media materials are developed and produced as planned; (b) the combination of campaign media are mobilized and coordinated as suggested; and (c) the campaign support personnel are trained accordingly. A management information system which monitors the campaign programme implementation can help provide campaign organizers with regular and up-to-date information for improving campaign management and performance. But the overall assessment on the impact or effects of campaign implementation has to be determined by a specific, and preferably built-in, campaign programme or summative evaluation.

Phase 8:   *Training of personnel*

While the need for training of campaign personnel is obvious, past campaign experiences have shown that in many instances training has been neglected. The implementation of most campaign activities requires new, different or more specific tasks and responsibilities to be added to the routine workload of the persons who have been mobilized for a campaign. However, in previous campaign activities, it has often been found that a considerable number of the campaign workers were not well-prepared for the specific tasks they were expected to perform. In some cases, they were not even adequately informed about the goals and specific issues of the campaign because their involvement with campaign activities was often conveyed through an administrative order. Such a bureaucratic approach to mobilizing campaign personnel, especially the front-line campaign workers, appears to be an inadequate means for ensuring that campaign workers are motivated and functionally capable of performing the additional tasks required of them to support the planned campaign activities.

Training sessions, or at least orientation meetings, for the various types and needs of campaign workers should be specifically planned and conducted before implementing a campaign. Such training activities are especially important if new or different skills and knowledge are required of campaign workers in performing their tasks effectively. Campaign workers should also be briefed adequately on how to integrate the campaign-related tasks with their routine work programme or activities, in addition to the purpose and value of their participation in the campaign programme.

Phase 9:   *Field implementation*

The most important element in ensuring that a campaign is implemented as planned is the appropriate monitoring and supervision of campaign workers' performance and the extension campaign delivery system. Such a task can be facilitated by a good management information system which is able to provide campaign organizers with rapid feedback on various important campaign activities and thus can help in readjusting or changing campaign strategies if considered necessary.

Another important task in campaign field implementation is the proper coordination of various activities which sometimes need to be carried out simultaneously. Coordination linkages must be carefully developed, especially if several agencies are involved in executing different aspects of the campaign activities. One of the most frequent problems found in campaign filed implementation is the untimely delivery, and often unavailability, of inputs or services required for the adoption of recommended technologies or action by the target beneficiaries who have been motivated and persuaded by the campaign. Such a problem may lead to a "frustration of rising expectations" among some members of the motivated target beneficiaries which in turn might undermine the credibility of the campaign recommendations and campaign workers.

Effective programme implementation also requires proper execution of activities within the estimated time period. A delay in one of the usually interdependent multi-media activities of a campaign may have chain-reaction effects. In planning the implementation of campaign activities, a realistic time estimate for the completion of an activity required to support the campaign should therefore be considered.

Phase 10: *Process documentation and summative evaluation*

Many people who are involved in rural development programmes have increasingly realized the need for, and the importance of conducting a summative evaluation. The main purpose of such an evaluation is to assess the performance, effects and impact of a campaign. Unlike formative evaluation which is normally conducted at the planning stage or early stages of campaign implementation, a summative evaluation is almost always conducted near or after the conclusion of a campaign. Whereas formative evaluation findings are often utilized to improve campaign strategy or performance during its implementation, the results of summative evaluation are normally used to determine whether the campaign has accomplished it objectives and if an improved or expanded campaign should be undertaken as a follow-up programme.

To ensure that a summative evaluation is conducted properly and that its results are relevant to the campaign objectives, summative evaluation activities should be considered as a built-in component and an integral part of the campaign process. The findings of such a summative evaluation should be used as inputs to formulate new or improved campaign objectives or to help set up new baselines or benchmarks for future campaigns of a similar nature. As can be seen in Figure 2-1, it is the information feedback resulting from the summative evaluation which completes the "loop" of the campaign planning process by feeding in relevant evaluation findings back to Phase 1 of such a process.

It is also useful to conduct a process documentation which points out critical issues and decision-making requirements in undertaking the SEC activities. Through a chronological description and analyses of successful or less-successful decision-making process made during planning, implementation, and management of the campaign, important lessons can be learned, and technical and management operation generalizations can be suggested, for future replications and expansion of similar activities.

Regardless of the outcome of a process documentation and summative evaluation exercise, the lessons learned (both positive and negative aspects) are valuable in designing or planning future campaigns which have similar objectives. Therefore, efforts to disseminate and share summative evaluation results of a strategic extension campaign (SEC) among extension planners, managers, and trainers should be encouraged and pursued. The publication of this book, for instance, is an attempt to share the experiences and results of SEC activities.

3. SEC METHODOLOGY APPLICATIONS

In this section, important aspects of the field implementation of a Strategic Extension Campaign (SEC), and the requirements for its replications to support various agricultural development, in different countries will be described. Examples on SEC replications, especially through building up of cadres of SEC planners and trainers, are also provided.

3.1. SEC Implementation Through FAO Projects

The first systematic effort in applying the SEC process and methodology was carried out in Bangladesh by FAO/UNDP project BGD/79/034 "Strengthening Agricultural Extension Service" which provided technical assistance to the Department of Agricultural Extension (DOAE). The work started in early 1982 with experimental campaigns on a small-scale basis covering three topics : (a) Methods of Seeds Storage, (b) Promoting Potato Consumption, and (c) Proper Processing of Khesari (a type of legume). In mid-1982, this project was requested to assist in the planning of a national campaign on Rodent Control by DOAE and the GTZ-funded project "Bangladesh-German Plant Protection Programme" (BGPPP). After more than four months of detailed planning and preparations, the National Campaign on Rodent Control was launched in January 1983, during the winter crop season, with wheat farmers as target beneficiaries. Empirical summative evaluation findings from the Information Recall & Impact Survey (IRIS) and also Field Damage Assessment (FDA) studies indicated that the campaign had accomplished positive results with significant economic benefits (see Section 4.1). The Bangladesh Government thus repeated the National Rodent

Control Campaign in 1984, and expanded its scope to cover all farmers in the country, and the 1984 campaign was as successful as that of 1983.

3.1.1

Examples of SEC Replications and Implementation Arrangements

Initially, the replications of SEC was limited to several Asian countries, and it can be attributed, to a large extent, to the adoption of the SEC method as one of the extension and training strategies of an FAO Inter-Country Programme on Integrated Pest Management (IPM) in Rice in South and Southeast Asia (projects GCP/RAS/101/NET, GCP/RAS/092/AUL, and GCP/RAS/108/AGF), hereafter referred to as the IPM project. The SEC method was integrated into this IPM project in 1984, towards the end of its first phase, when the remaining project budget was less than US$ 0.5 million.

The introduction of SEC in Malaysia, the Philippines, Thailand and Sri Lanka during 1985-1987, as the project's main IPM extension activities, produced concrete and tangible results. It also helped in creating an awareness of IPM benefits and a demand among farmers for IPM field training, as well as in mobilizing the planners and managers of those countries' national agricultural extension services to provide more specific IPM training and support materials for their field extension workers. Realizing the importance and cost-effectiveness of the extension & training component to support its programme, the IPM project allocated an estimated 55 percent of its Phase II total budget/resources of about US$ 12 million (for 1988-1992) for such a component.

Based on FAO experiences in planning and conducting SEC programmes in Bangladesh, Malaysia, Sri Lanka, Thailand and the Philippines, SEC was also adopted as the basic method for an FAO inter-regional project funded by UNFPA. The project which started in 1986 was on "Strategic Integration of Population Education into the Agricultural Extension Service", hereafter referred to as the PEDAEX project, with pilot activities in eight countries in Asia, Africa, the Near East and Caribbean/Latin America. As these pilot activities were successfully completed in 1992, nation-wide PEDAEX replications using the SEC method are now being implemented as a national project in Kenya, the Philippines, Uganda, and other national PEDAEX projects are planned for Jamaica, Morocco, Tunisia, Honduras, Rwanda etc.

In Africa and the Near East, SEC replications at the country level in most cases have been implemented within the context of FAO's field projects' extension and training activities. For instance, FAO/UNDP project ZAM/88/021 and TCP/ZAM/2254 project in Zambia utilized a standard SEC method which was evaluated positively by a UNDP evaluation team in 1992, and additional funding was recommended and approved by UNDP for a Phase II of project ZAM/88/021 (new project ZAM/92/022). In Thailand, starting in early 1993 the TCP/THA/2252 project has also utilised a SEC standard method in conducting its extension and training activities to support the Royal Thai programme on temperate-zone fruit crops cultivation. In French-speaking countries, the SEC or also called the Campagne Intensive de Vulgarisation pour l'Introduction d'un Thème (CIVIT) activities have been incorporated into FAO field projects, among others, in Burundi (BDI/85/011), in Guinea (GUI/86/004; GUI/86/012; and GUI/87/015), in Burkina Faso (BKF/89/018), in Rwanda (TCP/RWA/8958; and RWA/89/007), and in Tunisia (TUN/86/020).

Experiences from these countries indicate that the success of such SEC implementation depended on the degree or level of integration between the project activities and the programmes of the national agricultural extension service. In addition, positive results can more likely be obtained, if the complete SEC process is implemented, rather than implementing it partially or some of its operational phases only.

3.2. Institutionalization Problems : Neglect of SEC Training

While the 1983 and 1984 Rodent Control Campaigns in Bangladesh by themselves were quite successful, the institutionalization of SEC process and methodology in Bangladesh was less successful. In retrospect, one of the weaknesses of the SEC on Rodent Control in Bangladesh was that the staff of DOAE had not been specifically trained on SEC, although they were involved in campaign planning and implementation. The campaign activity was essentially a result of a "fire-brigade action" in response to an unplanned programme requested by high-level management of the Ministry of Agriculture. There was no time to provide structured and comprehensive training on SEC process and methodology to a core-group of extension planners/managers, subject-matter specialists, trainers, and communication support personnel. The campaign strategy and plan, its messages and multi-media materials, and management plan were developed mainly by a staff member of FAO project BGD/79/034, Dr. Ronny Adhikarya, assisted by Mr. Heimo Posamentier, a Rodent Biologist of the BGPPP, both international experts, and a few (about five) of their national counterparts. Such an ad-hoc and informal "on-the-job" training for a small group of national counterparts proved to be insufficient to enable them to conduct SEC replications without further assistance from the FAO and BGPPP projects.

It was regrettable that no SEC training, even for extension trainers and planners of DOAE, was conducted. One of the resource persons, Dr. Adhikarya, took up an FAO assignment in Rome, Italy after the completion of the 1983 Campaign and Mr. Posamentier left Bangladesh soon after the 1984 Campaign. Hence, they did not have the opportunity to organise SEC training for a care-group of extension planners and trainers to enable them to replicate SEC.

3.3.
Planned Replications of SEC : Building Cadres of SEC Planners & Trainers

Having learned from the lessons and experiences of SEC activities in Bangladesh, a more systematic and well-planned approach was adopted in introducing SEC process and methodology in Malaysia. In collaboration with the FAO's Inter-Country Programme on Integrated Pest Management in Rice in South and Southeast Asia (projects GCP/RAS/101/NET, GCP/RAS/092/AUL, and GCP/RAS/108/AGF), the FAO's Agricultural Education and Extension Service (ESHE) provided technical assistance to Malaysia's Department of Agriculture (DOA) in applying the SEC method. ESHE's staff member, Dr. R. Adhikarya, in 1985 developed the SEC training strategies, plans and materials, and organized five SEC skills-oriented workshops in Malaysia for 34 DOA staff and five staff members from two Malaysian universities. In addition, two persons from the Extension Division of the Department of Agriculture in Sri Lanka participated in these workshops.

The trainees from the DOA were extension officers, subject-matter specialists, trainers, and communication support officers who had seldom worked together, and had never been trained together on the same subject. Many of the 34 trainees participated in all of the five workshops. Due to some unavoidable problems, such as maternity/sick leave, duty transfer, etc., not all of them could not participate in all five workshops. The topics of the five workshops were the same as those shown in Fig. 2-2 (with the exception of Step 12).

In order to increase learning experience and ensure practicality of SEC training, the workshops were designed as a structured "on-the-job" training programme dealing with actual data, information, problems and solutions in planning, implementing and managing SEC activities. In 1985, Malaysia's Department of Agriculture had given a high priority to rat control, especially in Penang State. It requested FAO to focus its strategic extension campaign on rat control. Thus, a KAP survey on rat control in Penang State was carried out, and whose results were used as inputs for the SEC workshops. Based on the KAP survey results, workshop participants identified the rat control problems, formulated the rat control campaign objectives, developed the campaign strategy and management plans, and designed, developed, pretested and packaged the campaign messages and multi-media materials.

The above mentioned "learning-by-doing" SEC training method is a "killing two birds with one stone" activity which is pragmatic and result-oriented. Such an approach resulted not only in having well-trained personnel, but also actual campaign strategy and management plans, and pretested multi-media materials. In addition, it produced a group of SEC planners, managers and trainers who are highly motivated and dedicated in implementing real campaign activities whose strategies, plans and materials had been conceived and developed by themselves, and not produced and imposed by others. Unlike the Bangladesh experience, the success of Malaysia's Rat Control Campaign was not limited to achieving its stated objectives. More importantly, the SEC process and methodology has been replicated for other agricultural technologies, and by other institutions. As described in Section 4.2, there is considerable evidence to indicate that since 1987, the SEC method has been adopted as an integral part of Malaysia's agricultural extension service's programme and activities.

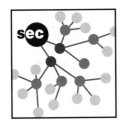

3.4.
Facilitating the Multiplier Effects

As will be explained in greater detail in Section 4, in addition to Bangladesh and Malaysia (Dept. of Agriculture), in Asia, FAO has assisted Sri Lanka, Thailand, the Philippines, China, and Malaysia (the Muda Agricultural Development Authority) in replicating the SEC process. In other regions, the SEC process has also been replicated in its entirety or partially, such as in Jamaica, Honduras, Zambia, Uganda, Liberia, Malawi, Kenya, Rwanda, Benin, Burundi, Burkina Faso, Guinea, Morocco, Tunisia, etc.

3.4.1   *Training of SEC "Master Trainers"*

One of the most effective strategies which resulted in the "snow-balling" of SEC activities in many countries is the use of SEC trained personnel as "master trainers" or resource persons for replication of SEC process in their own country (for other agricultural agencies or technologies) and/or in other countries. The use of such persons as SEC multiplier agents has been effectively demonstrated in Asia. In addition to replicating the SEC process within their own local institutions, many participants who attended the first series of SEC workshops during 1985-87 in Asia have served as SEC trainers or resource persons in replicating the SEC process, among others, in Thailand, Sri Lanka, China, the Philippines, Malawi, Zambia, Uganda, and Jamaica. A list of selected SEC training participants who later have served as SEC resource persons to assist FAO extension and training projects around the world is shown in Figure 3-1.

3.4.2   *Spreading the "SEC-Fever" Beyond Asia*

Based on the experience gained in Asia, FAO's Agricultural Education and Extension Service (ESHE) replicated the SEC process in Africa.

FIGURE 3-1

Selected List of Trained SEC Resource Persons
(As of 1 July 1994)

SEC Resource Persons	SEC Training/Workshop Participated in			Served as SEC Resource Person* in the following FAO project(s)		
	Year	Country	Project No./Activity	Year	Country	Project No./Activity
R. Adhikarya (FAO)	1981-83	Bangladesh	BGD/79/034	1981-83	Bangladesh	BGD/79/034
	1985-89	Malaysia	GCP/RAS/101/NET	1985-89	Malaysia	GCP/RAS/101/NET
	1987-88	Thailand	GCP/RAS/101/NET	1987-88	Thailand	GCP/RAS/101/NET
	1987	Liberia	FAO Reg. Prog.	1987	Liberia	FAO Reg. Prog.
	1991	Malaysia	TMD review meeting	1988-89	Sri Lanka**	GCP/RAS/108/AGF
	1993	Thailand	Use of SEC-TM wksp	1989-90	Philippines**	GCP/RAS/092/AUL
	1994	Malaysia	Use of KAP-TM wksp	1993-94	Thailand**	TCP/THA/2252 TMD°° Coordinator
A.S.M. Noor (Malaysian)	1985-86	Malaysia	GCP/RAS/101/NET	1987-88	Thailand	GCP/RAS/101/NET
	1991	Malaysia	TMD review meeting	1987-89	Malaysia	GCP/RAS/101/NET
	1993	Thailand	Use of SEC-TM wksp	1989-90	Philippines	GCP/RAS/092/AUL
	1994	Malaysia	Use of KAP-TM wksp	1989	Malawi	MLW/88/P04 TMD°° Team Member
A. Hamzah (Malaysian)	1985	Malaysia	GCP/RAS/101/NET	1986	Malaysia	GCP/RAS/101/NET TMD°° Associate
	1991	Malaysia	TMD review meeting			
A.B. Othman (Ms.) (Malaysian)	1985-86	Malaysia	GCP/RAS/101/NET	1987-88	Thailand	GCP/RAS/101/NET
	1991	Malaysia	TMD review meeting	1987-89	Malaysia	GCP/RAS/101/NET
				1988-89	Sri Lanka	GCP/RAS/108/AGF TMD°° Associate
K.L. Jayatissa (Sri Lankan)	1985	Malaysia	GCP/RAS/101/NET	1988-89	Sri Lanka	GCP/RAS/108/AGF TMD°° Associate
	1991	Malaysia	TMD review meeting			
Y.L. Khor (Ms.) (Malaysian)	1986	Malaysia	GCP/RAS/101/NET	1987-89	Malaysia	GCP/RAS/101/NET
	1991	Malaysia	TMD review meeting	1988-89	Sri Lanka	GCP/RAS/101/NET
	1993	Thailand	Use of SEC-TM wksp	1989-90	Philippines	GCP/RAS/092/AUL
	1994	Malaysia	Use of KAP-TM wksp	1990	China	CPR/88/067 TMD°° Team Member
R. Mohamed (Malaysian)	1986	Malaysia	GCP/RAS/101/NET	1987-89	Malaysia	GCP/RAS/101/NET
	1991	Malaysia	TMD review meeting	1988-89	Sri Lanka	GCP/RAS/108/AGF
	1993	Thailand	Use of SEC-TM wksp	1990	Thailand	THA/89/P04
	1994	Malaysia	Use of KAP-TM wksp	1990	Jamaica	JAM/89/P02
				1991-92	Zambia	ZAM/88/P08
				1992	Uganda	UGA/88/P08
				1993-94	Thailand	TCP/THA/2252 TMD°° Team Member
J. Jallade (FAO)	1987	Liberia	FAO Reg. Prog.	1989	Burundi	BDI/85/011
	1988	Malaysia	GCP/RAS/101/NET	1989	Benin	BEN/87/011
	1988	Thailand	GCP/RAS/101/NET	1989	Tunisia	TUN/86/020
				1989-91	Rwanda	TCP/RWA/8958
				1990-91	Morocco	FAO Reg. Prog.

* Involved in a part of, or the complete, SEC process; ** Initiated and planned the overall SEC activities only.
°° Participated in SEC Training Module Development (TMD) preparation, writing, testing, and/or production activities
TM wksp=Training Module workshop.

FIGURE 3-1 (CONTINUED)

Selected List of Trained SEC Resource Persons
(As of 1 July 1994)

SEC Resource Persons	SEC Training/Workshop Participated in			Served as SEC Resource Person in the following FAO project(s)		
	Year	Country	Project No. / Activity	Year	Country	Project No./ Activity
M. Escalada (Ms.) (Filipina)	1987 1991 1993 1994	Thailand Malaysia Thailand Malaysia	GCP/RAS/101/NET TMD review meeting Use of SEC-TM wksp Use of KAP-TM wksp	1987-88 1988-89 1989-90 1992-93	Thailand Malaysia Philippines Philippines	GCP/RAS/101/NET GCP/RAS/101/NET GCP/RAS/092/AUL PHI/90/P28 TMD°° Team Member
N.K. Ho (Malaysian)	1987 1988	Malaysia Malaysia	GCP/RAS/101/NET GCP/RAS/101/NET	1988-89	Malaysia	GCP/RAS/101/NET
P. Piradej (Thai)	1981 1987	Bangladesh Thailand	BGD/79/034 GCP/RAS/101/NET	1987-88 1990 1990 1993	Thailand China Thailand Thailand	GCP/RAS/101/NET CPR/88/067 THA/89/P04 TCP/THA/2252
B. Amoako-Atta (Ghanaian)	1987 1991	Liberia Malaysia	FAO Reg. Prog. TMD review meeting	1988 1989	Zambia Malawi	ZAM/86/003 MAL/88/P04 TMD°° Associate
Y. Ayele (Ethiopian)	1987	Liberia	FAO Reg. Prog.	1987-88	Liberia	TCP/LIR/5754
R. Soemarsono (Indonesian)	1988 1988	Malaysia Thailand	GCP/RAS/101/NET GCP/RAS/101/NET	1988	Indonesia	TCP/INS/4515
T. Syafii (Ms.) (Indonesian)	1988 1988	Malaysia Thailand	GCP/RAS/101/NET GCP/RAS/101/NET	1988	Indonesia	TCP/INS/4515
E. Baier (FAO)	1989 1989 1990	Malawi Tunisia Morocco	MAL/88/P04 TUN//88/P02 MOR/88/P07	1990 1991	Kenya Rwanda	KEN/89/P01 RWA/88/P01
G. Sévin (Français)	1989	Tunisia	TUN/88/P02	1989 1990	Tunisia Morocco	TUN/88/P02 MOR/88/P07
A. Boufaroua (Tunisien)	1989	Rwanda	TCP/RWA/8958	1989-90	Tunisia	TUN/86/020
J. L. Michard (Français)	1989	Rwanda	TCP/RWA/8958	1989	Rwanda	TCP/RWA/8958
J.F. Gascon (FAO)	1989 1990-91	Rwanda Morocco	TCP/RWA/8958 FAO Reg. Pro.	1989-90	Rwanda	TCP/RWA/8958
J. Ngamijiyarémyé (Rwandais)	1989-90	Rwanda	TCP/RWA/8958	1991-93	Rwanda	RWA/89/007 (on 3 different topics)
B. Gasanguwa (Rwandais)	1989-90	Rwanda	TCP/RWA/8958	1991-93	Rwanda	RWA/89/007 (on 3 different topics)

* Involved in a part of, or the complete, SEC process; ** Initiated and planned the overall SEC activities only.
°° Participated in SEC Training Module Development (TMD) preparation, writing, testing, and/or production activities.
TM wksp=Training Module workshop

FIGURE 3-1 (CONTINUED)

Selected List of Trained SEC Resource Persons
(As of 1 July 1994)

SEC Resource Persons	SEC Training/Workshop Participated in			Served as SEC Resource Person in the following FAO project(s)		
	Year	Country	Project No./ Activity	Year	Country	Project No./ Activity
J.Mukumana (Mme) (Rwandaise)	1989-90	Rwanda	TCP/RWA/8958§	1991-93	Rwanda	RWA/89/007 (on 3 different topics)
G. Masabo (Burundais)	1989	Rwanda	TCP/RWA/8958	1989-90	Burundi	BDI/85/011
P. Nigagura (Burundais)	1989	Rwanda	TCP/RWA/8958	1989-90	Burundi	BDI/85/011
R. Bonkoungou (Burkinabè)	1989 1989 1990 1991	Rwanda Burundi Morocco Morocco	TCP/RWA/8958 BDI/85/011 FAO Reg. Pro. FAO Reg. Pro.	1992-93	Burkina Faso	BKF/89/018
S. Sanoh (Guinéen)	1989 1989 1990 1991	Rwanda Burundi Morocco Morocco	TCP/RWA/8958 BDI/85/011 FAO Reg. Pro. FAO Reg. Pro.	1991-92 1991-92 1991-92	Guinea Guinea Guinea	GUI/86/004 GUI/86/012 GUI/86/017
M. Malère (Français)	1990 1990 1991	Rwanda Morocco Morocco	TCP/RWA/8958 FAO Reg. Pro. FAO Reg. Pro.	1991-92 1991-92 1991-92	Guinea Guinea Guinea	GUI/86/004 GUI/86/012 GUI/86/017
B. Thiam (Mauritanien)	1990 1991 1991-92	Morocco Morocco Guinea	FAO Reg. Pro. FAO Reg. Pro. GUI/86/004	1991-92	Mauritania	MAU/88/002
A. Le Magadoux (Mme) (Français)	1990 1991	Morocco Morocco	FAO Reg. Pro. FAO Reg. Pro.	1990 1991	Rwanda Guinea	TCP/RWA/8958 GUI/86/004
L. Larribe (FAO)	1990 1991	Morocco Morocco	FAO Reg. Pro. FAO Reg. Pro.	1993	Zaire	ZAI/88/006
C.H. Teoh (Australian)	1990 1991 1993 1994	Thailand Malaysia Thailand Malaysia	THA/89/P04 TMD review meeting Use of SEC-TM wksp Use of KAP-TM wksp			TMD°° Team Member
S. Sudsawasd (Ms.) (Thai)	1990 1991 1993 1994	Thailand Malaysia Thailand Malaysia	THA/89/P04 TMD review meeting Use of SEC-TM wksp Use of KAP-TM wksp	1990-91 1993	Thailand Thailand	THA/89/P04 TCP/THA/2252 TMD°° Associate
K. Jarinto (Thai)	1990 1991	Thailand Malaysia	THA/89/P04 TMD review meeting	1990-91	Thailand	THA/89/P04 TMD°° Associate
K. Sondi (Zairois)	1991-92	Guinea	GUI/86/004	1993	Zaire	ZAI/88/006

* Involved in a part of, or the complete, SEC process; ** Initiated and planned the overall SEC activities only.
°° Participated in SEC Training Module Development (TMD) preparation, writing, testing, and/or production activities.
TM wksp=Training Module workshop

FIGURE 3-1 (CONTINUED)

Selected List of Trained SEC Resource Persons
(As of 1 July 1994)

SEC Resource Persons	SEC Training/Workshop Participated in			Served as SEC Resource Person in the following FAO project(s)		
	Year	Country	Project No./ Activity	Year	Country	Project No./ Activity
J. Mesfin (Ethiopian)	1991-92 1993 1994	Zambia Thailand Malaysia	ZAM/88/021 Use of SEC-TM wksp Use of KAP-TM wksp	1992 1992-1993 1994	Zambia Zambia Tanzania	ZAM/88/021 TCP/ZAM/2254 URT/89/019
B. Katon (Ms.) (Filipina)	1993	Thailand	Use of SEC-TM wksp	1993-1994	Philippines	PHI/90/P29
M. Rikhana (Ms.) (Indonesian)	1993 1994	Thailand Malaysia	Use of SEC-TM wksp Use of KAP-TM wksp	1993-1994 1994	Indonesia Indonesia	FAO Reg.Prog.Act. INT/92/P95
X.Y. Li (Chinese)	1993 1994	Thailand Malaysia	Use of SEC-TM wksp Use of KAP-TM wksp			
A. Halim (Bangladeshi)	1993 1994	Thailand Malaysia	Use of SEC-TM wksp Use of KAP-TM wksp	1994 1994	Bangladesh Bangladesh	FAO Reg.Prog.Act. INT/92/P95
A. Apinantara (Thai)	1993	Thailand	Use of SEC-TM wksp			
C. Lertchalolarn (Thai)	1993	Thailand	Use of SEC-TM wksp			
D. Rerkrai (Thai)	1993	Thailand	Use of SEC-TM wksp			
N. Sompong (Thai)	1993	Thailand	Use of SEC-TM wksp			
W. Boonyatharokul (Thai)	1993	Thailand	Use of SEC-TM wksp			
S. Seesang (Thai)	1993 1994	Thailand Malaysia	Use of SEC-TM wksp Use of KAP-TM wksp			
T. Rungrawd (Thai)	1993 1994	Thailand Malaysia	Use of SEC-TM wksp Use of KAP-TM wksp	1994 1994	Thailand Thailand	FAO Reg.Prog.Act. INT/92/P95
N. Pattanapongsa (Thai)	1993 1994	Thailand Malaysia	Use of SEC-TM wksp Use of KAP-TM wksp	1993-94 1994	Thailand Thailand	TCP/THA/2252 INT/92/P95
J. Mbindjo (Kenyan)	1994	Malaysia	Use of KAP-TM wksp	1994	Kenya	INT/92/P95
M.C.B. Oliveira (Ms.) (Brazilian)	1994	Malaysia	Use of KAP-TM wksp			

* Involved in a part of, or the complete, SEC process; ** Initiated and planned the overall SEC activities only.
°° Participated in SEC Training Module Development (TMD) preparation, writing, testing, and/or production activities.
TM wksp=Training Module workshop

Small experimental SEC activities were conducted in Zambia through project TCP/ZAM/6658 whose extension training expert was technically backstopped by ESHE. The campaign topics for which campaign materials were developed and produced included : (a) Horticulture Production, and (b) Cattle Management. However, the first major intensive effort in replicating the SEC process in Africa which included a comprehensive KAP survey and staff training was conducted in Liberia in 1987 by ESHE, supported by its Regular Programme funds. The topic selected for the campaign in Liberia was on Line Sowing Method of Rice Cultivation. An ESHE's staff member, Dr. R. Adhikarya, who initiated the SEC activities in Asia, conducted the first African SEC training programme in Liberia whose participants developed a campaign strategy plan and multi-media support materials. With the assistance of an FAO's extension project TCP/LIR/5754, these materials were later pretested and mass-produced for use by Liberia's extension workers. Most of the 30 participants of this SEC training were from the Liberia's Ministry of Agriculture but other participants included staff members of the West African Rice Development Association (WARDA), the University of Liberia, Radio Liberia and resident staff of GTZ and IIR/USAID. Another ESHE staff member, Mr. Jacques Jallade, a technical officer responsible for the extension component of most FAO's field projects in Francophone Africa, also participated in the SEC training in Liberia, and has since been very instrumental in spreading the "SEC-fever" and replicating SEC method through FAO field projects in French-speaking countries in Africa.

Since the first African SEC training (conducted in Liberia, in April 1987), a number of SEC activities have been undertaken in various parts of Africa, the Near East and the Caribbean. Under the active coordination and leadership of Mr. J. Jallade of FAO/ESHE, who subsequently also attended other SEC workshops in Thailand and Malaysia in 1988, the SEC process and method (called Campagne Intensive de Vulgarisation pour l'Introduction d'un Thème or CIVIT) have been replicated in French-speaking countries in Africa and the Near East. As of July 1994, many SEC-related activities have been undertaken through FAO projects and FAO's Regular programme (incl. TCP) activities in French-speaking countries such as: Burundi, Rwanda, Guinea, Burkina Faso, Tunisia, Morocco, and other SEC activities are also planned for Zaire and Mauritania.

One of the participants who attended the Liberia SEC workshop and who was involved in conducting the KAP Survey on Line Sowing in Liberia, Dr. Amoako-Atta from Ghana, served as an FAO consultant in Zambia in 1989 to conduct SEC activities for Department of Agriculture. He also served as a resource person in another SEC workshop held in 1989 by FAO/ESHE to support an FAO/UNFPA project in Malawi. This Malawi workshop, with population education as the topic, was conducted and coordinated by a Malaysian, Mr. A. Saffian Mohd. Noor, who participated in the SEC training in Malaysia and had assisted in SEC replications in Thailand, Sri Lanka and the Philippines. A participant (Mr. Muyaya) in the Malawi SEC activities, also later served as a member of a mission to evaluate an FAO's SEC activities on population education in Kenya. Another resource person, Dr. R. Mohamed who participated in SEC workshop/training activities in Malaysia, Thailand, and Sri Lanka, has assisted with SEC planning, multi-media materials development, and training activities for FAO projects in Zambia, Uganda and Jamaica. Figure 3.1 shows a list of SEC resource persons who have been involved in various FAO-sponsored extension and training activities.

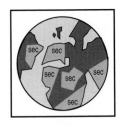

3.5. Strategy of SEC Implementation at Country Level

Another strategy which FAO has utilized in the SEC replication process is to work with a national extension service through the activities of an on-going FAO's field project or its Regular Programme which provides technical assistance in a given country. In order to "promote" the SEC approach, its tangible results and concrete benefits must be demonstrated to important national agricultural policy and decision-makers. It is not adequate to discuss only the SEC institutionalization ideas or approach in abstract terms with them. As such, the results of at least a pilot SEC programme conducted in that country must be well-documented. Convincing evidence of its usefulness and benefits, including actual SEC strategies and plans, multi-media materials, trained personnel, and positive evaluation findings, must be shown. In order to demonstrate such a case, the complete SEC process and activities should be initiated with FAO assistance, preferably through an on-going FAO field project. It should be completed in a relatively short time period (12-18 months). Without such initial support from an FAO project, it would be difficult to ensure the necessary follow-up actions for the implementation of initial SEC field activities, as well as the availability of operational funds. The following are the strategy guidelines for SEC implementation in a given country :

1   It should be integrated as part of a national extension programme

2   It should be executed initially as part of an FAO project

3   Government commitment to support field implementation activities of an SEC demonstration programme for a period of at least 18 months should be obtained

4   The SEC topic should focus on a specific extension problem considered as a high priority by the Ministry of Agriculture

5   A core-group of at least 25 persons representing extension planners/managers, trainers, subject-matter specialists, communication support staff, etc. should be assigned by the Government to participate in all SEC activities, including training and field implementation

6   A formative and summative evaluation of an SEC demonstration programme should be conducted, and supplemented with printed, audio and/or visual documentation of its process and results

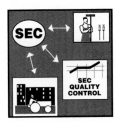

3.6.
Facilitating SEC Institutionalization and Quality Control

| 3.6.1 | *Develop SEC Training Curricula and Materials* |

One of the challenges in replicating SEC process in different countries for various technology transfer purposes is how to maintain a high degree of quality control in applying the SEC methodology. The SEC experience thus far suggests that training of extension staff (especially the trainers) on the skills and techniques of SEC is of critical importance in applying effectively the SEC methodology, and in institutionalizing SEC as a basic method of an agricultural extension service. The FAO's Agricultural Education and Extension Service (ESHE) has been developing standardized SEC process documentation procedures as well as preparing and testing SEC training modules and materials for use in training of extension trainers activities.

Training curricula and support materials related to specific SEC skills/techniques, such as Knowledge, Attitude and Practice (KAP)/baseline survey, Strategic Extension Planning and Message Design, Multi-media Materials Development and Pretesting, Campaign Management and Monitoring, Summative Evaluation, etc. have been developed and pretested for use to train SEC trainers, potential resource persons and/or consultants. Some SEC training materials have also been adapted, and translated into French, for use in African Francophone countries to conduct the Strategic Extension Campaign activities, or the Campagne Intensive de Vulgarisation pour l'Introduction d'un Thème (CIVIT).

3.6.2   *Develop and Utilize a Core-Group of SEC Resource-Persons*

Training programmes for SEC resource persons and trainers in countries which plan to use the SEC or CIVIT methodology have also been conducted. Two regional workshops on SEC methodology were conducted in 1990 and 1991 in Morocco by FAO and attended by 20 participants from 10 French-speaking countries sponsored by UNDP, World Bank, FED, CCCE, etc. A SEC training module development review meeting was conducted in Malaysia in July 1991 for 12 SEC trainers from 6 English-speaking countries. Another SEC training materials development workshop for 20 SEC trainers from 10 countries was organized in 1993 in Thailand.

As already described in Section 3.4., competent and experienced SEC trainers or planners from a given country should also be mobilized and given the opportunity to widen and share his/her SEC experiences by serving as SEC resource persons in other countries or regions. The wide and rapid spread of the SEC methodology could not have been possible without applying the training of trainers strategy as shown in Fig. 3-1. Such a strategy in developing and utilizing specifically trained human resources is critical to the efforts of capacity and institution building of national extension system and can contribute significantly to ensuring its programme sustainability. These SEC resource persons have also been involved in the development of standardized or generic SEC training curricula and materials as can be seen in Fig. 3-1, so that their practical experiences are reflected and included in the SEC training modules and/or materials.

4. SEC PROGRAMMES: HIGHLIGHTS OF RESULTS

In this chapter, some of FAO's assisted SEC programmes will be described, and their important results will be highlighted. Examples on how SEC programmes were planned, implemented and evaluated within the context of FAO projects and incorporated into activities of national agricultural extension services will also be provided. Since the first SEC programme was implemented by an FAO project in Bangladesh in 1983, many other SEC replications had been carried out in different countries. Not all, however, had conducted the entire SEC process. Some activities were not completed because no follow-up support was made available. Others only conducted parts or elements of the SEC process which had been missing or weak (such as KAP survey, pretesting of extension materials, strategy development and planning, etc.), in order to enhance or strengthen their extension programmes. Figure 4-1 shows an inventory-analysis of FAO-supported SEC activities conducted in different countries.

FIGURE 4-1

Inventory-Analysis of FAO-Supported SEC Activities: as of December 31, 1993

STRATEGIC EXTENSION CAMPAIGN (SEC) ELEMENTS

COUNTRY / SEC Topic(s)	KAP SURVEY	WORKSHOPS* 1	2	3	4	5	STRATEGY PLANNING	MULTI-MEDIA SUPPORT MATERIALS — Field Pretesting	Revision/ Production	Distribution to Users	MANAGEMENT PLANNING	STAFF TRAINING	FIELD IMPLEMENTATION	EVALUATION — MMS	IRIS	FIELD PROJECT SUPPORT	TIME PERIOD
BANGLADESH:																	
Promoting potato consumption			C	C			C	C	C	C		C	C	C		BGD/79/034	1982
Seed storage methods			C	C			C	C	C	C		C	C	C		BGD/79/034	1982
Processing of "Khesari"			C	C			C	C	C	C		C	C			BGD/79/034	1982
Rodent control 1983							C	C	C	C	C	C	C	C	C	BGD/79/034	1983
Rodent control 1984							C	C	C	C	C	C	C	C	C	BGD/79/034	1984
MALAYSIA:																	
Rat control	C		C	C	C	C	C	C	C	C	C	C	C	C	C	GCP/RAS/101/NET	1985/86
Integrated weed management	C		C	C	C	C	C	C	C	C	C	C	C	C	C	GCP/RAS/101/NET	1988/89
PHILIPPINES :																	
Coconut replanting	C	C														TCP/PHI//4405	1985
Integrated pest control	C		C	C	C		C	C	C	C	C		C		C	GCP/RAS/092/AUL	1986
Golden snail (Kuhol) control	C		C	C	C	C	C	C	C	C	C	C	C			GCP/RAS/092/AUL	1988/89
Population education	C		C	C	C		C	C	C	C	C	P	P	P	P	PHI/90/P28	1992/94
Population education	C		C	C	C		C	C	C	C	C	P	P	P	P	PHI/90/P29	1992/94
LIBERIA:																	
Line sowing method for rice cultivation	C		C	C			C	C	C	C	C	C	C			GCP/LIR/5754	1987

Notes: C = Completed/Conducted
P = in Progress
MMS = Management Monitoring Survey
IRIS = Information Recall & Impact Survey

* SEC Workshop : 1 = Purpose & Methodology of KAP Survey
2 = Strategy Planning, Message Design, and Prototype Materials Development
3 = Methods of Pretesting Multi-Media Materials
4 = Campaign Management Planning
5 = Methods of Campaign Evaluation & Management Monitoring Survey

FIGURE 4-1 CONTINUED

Inventory-Analysis of FAO-Supported SEC Activities: as of Dec. 31, 1993 (continued)

STRATEGIC EXTENSION CAMPAIGN (SEC) ELEMENTS

COUNTRY / SEC Topic(s)	KAP SURVEY	WORKSHOPS* 1	2	3	4	5	STRATEGY PLANNING	MULTI-MEDIA SUPPORT MATERIALS Field Pretesting	Revision/ Production	Distribution to Users	MANAGEMENT PLANNING	STAFF TRAINING	FIELD IMPLEMENTATION	EVALUATION MMS	IRIS	FIELD PROJECT SUPPORT	TIME PERIOD
THAILAND:																	
Pest surveillance system	C	C	C	C	C	C	C	C	C	C	C	C	C	C	C	GCP/RAS/101/NET	1987/88
Population education	C	C	C	C	C		C	C	C	C	C	C	C	C	C	THA/89/P04	1990/91
Temperate-zone fruit crops cultivation	C		C	C	C	P	C	C	C	C	C	C	C	P	P	TCP/THA/2252	1993/94
SRI LANKA:																	
Integrated pest management	C	C	C	C	C	C	C	C	C	C	C	C	C	C		GCP/RAS/108/AGF	1987/89
ZAMBIA:																	
Horticulture production	C	C	C	C	C		C	C	C	C	C	C	C			TCP/ZAM/6658	1987/88
Cattle management	C	C	C	C	C		C	C	C	C	C	C	C			TCP/ZAM/6658	1987/88
Maize prod. & post harvesting	C	C	C	C	C		C	C	C	C	C	C	C			ZAM/86/003	1988/89
Maize cultivation	C	C	C	C	C	C	C	C	C	C	C	C	C	C	C	ZAM/88/021	1991/92
BENIN:																	
Ploughing with drought animal power	C	C														BEN/87/011	1989/91
BURUNDI:																	
Tick-borne disease control	C	C	C	C	C	C	C	C	C	C	C	C	C	C		BDI/85/011	1989/90
HONDURAS:																	
Population education	C		C	C	C		C	C	C	C	C	C	C	C	C	HON/88/P05	1989/91
MALAWI:																	
Population education	C		C	C	C		C	C	C	C	C	C	C	C	C	MLW/88/P04	1989/91
MOROCCO:																	
Population education	C		C	C	C		C	C	C	C	C	C	C	C	C	MOR/88/P07	1989/91

FIGURE 4-1 (CONTINUED)

Inventory-Analysis of FAO-Supported SEC Activities: as of Dec. 31, 1993 (continued)

STRATEGIC EXTENSION CAMPAIGN (SEC) ELEMENTS

COUNTRY / SEC Topic(s)	KAP SURVEY	WORKSHOPS * 1	2	3	4	5	STRATEGY PLANNING	MULTI-MEDIA SUPPORT MATERIALS Field Pretesting	Revision/ Production	Distribution to Users	MANAGE-MENT PLANNING	STAFF TRAINING	FIELD IMPLEMEN-TATION	EVALUATION MMS	IRIS	FIELD PROJECT SUPPORT	TIME PERIOD
RWANDA:																	
Bean seeds coating	C	C	C	C	C		C	C	C	C	C	C	C	C	C	TCP/ZAM/6658	1989/91
Population education	C	C	C	C	C		C	C	C	C	C	C	C	C	C	RWA/88/P02	1990/92
Internal parasite control for cattle	C		C	C	C	C	C	C	C	C	C	C	C	C	C	RWA/89/007	1992/93
Population education	C	C	C	C	C	C	C	C	C	C	C	C	C	C		RWA/89/007	1992/93
TUNISIA:																	
Contour tillage	C	C	C	C	C	C	C	C	C	C	C	C	C	C		TUN/86/020	1989/90
Population education	C		C	C	C		C	C	C	C	C	C	C	C	C	TUN/88/P02	1989/91
JAMAICA:																	
Population education	C		C	C	C		C	C	C	C	C	C	C	C	C	JAM/89/P02	1990/91
KENYA:																	
Population education	C		C	C	C		C	C	C	C	C	C	C	C	C	KEN/89/P01	1990/91
GUINEA:																	
Mechanical & chemical control of termites	C	C	C	C	C	C	C	C	C	C	C	C	C			GUI/86/004	1991/92
Fodder stocks	C	C	C	C	C	C	C	C	C	C	C	C	C			GUI/86/012	1991/92
Composting	C		C	C	C	C	C	C	C	C	C	C	C			GUI/87/015	1991/92
BURKINA FASO:																	
Composting	C	C	C	C	C	C	C	C	C	C	C	C	C	P	P	BKF/89/018	1992/93
UGANDA:																	
Population education	C	C	C	C	P	P	C	C	C	C	P	P	P	P		UGA/88/P08	1992/93

4.1.
Bangladesh: the Rodent Control Campaigns

In 1983 and 1984, the Bangladesh's Department of Agricultural Extension (DOAE) in collaboration with Bangladesh-German Plant Protection Programme (BGPPP) and FAO/UNDP project (BGD/79/034), which provided the technical assistance, launched two nation-wide Rodent Control Campaign programmes. The 1983 Campaign was aimed at wheat farmers and its general objective was to increase their rat control practice from 10 percent to 25 percent in one year. The decision to focus the campaign on wheat crop was due to the higher rat-related damage in wheat fields as compared to other crops. In addition, wheat has become an increasingly important food crop in Bangladesh, earning it a high priority in agricultural policy and production. The Campaign was conducted during the winter crop season (Jan.-March, 1983). It covered 11 districts as its primary target area, which accounted for about 91 percent (1,500 acres) of the total wheat acreage, and about 90.7 percent of the total wheat production in Bangladesh. Due to some political considerations (i.e., "equity" issue, etc.), the other 10 districts in the country were considered as a "secondary" target area of the campaign. However, only minimal campaign inputs were provided to these "secondary" districts, and for evaluation purposes it could thus be considered as a "control" group.

The campaign's target beneficiaries were segmented according to their levels of knowledge, attitudes and practices concerning rat control. Positioning of motivational messages included the use of various appeals such as religious incentive, fear arousal, guilt feeling, as well as a ridicule appeal that served as a discussion point. The multi-media strategy plan for the 1983 Campaign is shown in Figure 4-2. Radio and posters were used to provide general information and motivational messages. Extensive training sessions were conducted to motivate and educate campaign workers on their specific tasks before the campaign. Interpersonal support was provided by extension workers who conducted small group discussions, field demonstrations and farmers training sessions. Teachers, school

Strategic Multi-Media Plan of the 1983 Rat Control Campaign

FARMERS IN 11 WHEAT GROWING DISTRICTS

FARMERS IN NON-WHEAT GROWING DISTRICTS

CAMPAIGN CENTER

Motivational Poster

Comic Sheet

Essay Competition

Leaflet

Instructional Poster

Training

Krishi Khata Magazine

Extension Newsletter

Radio: Rajshahi, Rangpur, Dhaka Stations (Spots, Jingles, Songs & Drama)

Television (Slide & Sound)

Campaign Inauguration Ceremony

Leaflet

Instructional Poster

Motivational Poster

Leaflet

Extension Workers

Secondary School Teachers

Extension Workers

Community Local Leaders

News Media

Extension Workers

Extension Workers

Pesticide & Agri. Inputs Dealers

Markets, Tea-stalls, Bus-stops & Mosques

Government Offices

Secondary School Children

Parents

Contact Farmers

Radio BGD

Newspaper, Magazine, Radio, Television Reports

Pesticide Dealers

Pesticide Clients

Pesticide & Agri. Inputs Dealers

Markets, Tea-stalls, Bus-stops & Mosques

Government Offices

Interpersonal/Group Discussion

Interpersonal/Group Discussion

T &V Meetings

Endorsement & Legitimization

Discussion

Discussion

Training

Training

Delivery Agents/Channels

Intermediaries

Distribution/display means or points

Source: Adhikarya & Posamentir (1987)

FIGURE 4-3 a

Various Inputs into the 1983 Campaign

Mass Communication Inputs

Districts	Motivational Posters	Instructional Posters	Comic Sheets	Leaflets	Radio	Television	Newspapers/ Magazines
Primary Targets (11 districts)	15,000 copies	15,000 copies	375,000 copies	100,000 copies	87.5 minutes scheduled air time (20 different programmes) and news reports	21 minutes scheduled air time (spread over 7 days) and news reports	Features, news reports and 4 advertisements on essay competition
Secondary Targets (10 districts)	10,000 copies	–	–	15,000 copies	Unplanned, but possible spill-over effects	Unplanned, but possible spill-over effects	Unplanned, but possible spill-over effects

Districts	Personal Communication/Influence Inputs					Special Events		Technology Inputs — Ready-Made Baits Through	
	Extension workers	School children's parents	Community/Local leaders	Pesticide/ Seed dealers	Pesticide/ Seed clients	Campaign inauguration ceremony	Essay competition for school children	Extension workers	Commercial retail outlets/ pesticides or seed dealers
Primary Targets (11 districts)	Yes	Yes	Yes	Yes	Yes	At national, districts and upazila levels	At national, districts, upazila and union levels	80,000 packets	At least 140,000 packets
Secondary Targets (10 districts)	Unplanned, but possible spill-over effects	–	–	–	–	–	Unplanned, but possible spill-over effects (due to newspaper advertisements)	–	Not monitored, but possible spill-over effects

FIGURE 4-3 b

Costs of the 1983 Campaign

Expenses for	Taka	%
Design, development and production of printed materials (including paper)	204,187	59
Newspaper advertisements	5,850	2
Campaign inauguration ceremony (national level)	18,000	5
Campaign inauguration ceremonies (district and upazila levels)	41,800	12
Incentives for essay competition	34,650	10
Distribution of campaign materials	6,000	2
Evaluation studies	33,000	10
Miscellaneous	4,700	1
Total	Taka 348,187 (US $ 17,409 equiv.)	

Note: 20 Taka = US $ 1 (1983 exchange rate)

*Extension campaign
planning and
pretesting of materials sessions*

Motivational poster

Instructional poster

Comic sheet

Logo before pre-testing

Logo after pre-testing

Motivational Poster

Samples of the Bangladesh rat control campaign printed materials.

dren and agricultural supply/inputs retailers were involved to augment the extension workers. For example, comics were distributed through rural schools by teachers and were taken home by school children to discuss with their parents, most of whom are illiterate. This component was complemented by an essay contest for school children whose topic was "My Parents' Problems and Experience in Fighting Rats". Family and/or community level discussion was generated as rural school children had to consult the rat problem issue with their parents to enable them to write the essay. Figure 4-3a shows the complete campaign inputs.

Encouraged by the 1983 Campaign results, DOAE repeated the Campaign in 1984. Its scope was expanded to reach all farmers, thus not limited to wheat farmers. While the 1983 campaign succeeded in increasing the number of farmers who practise rat control, evaluation data indicated that the increase was accounted mainly by large farmers (cultivating more than 5 acres) and medium farmers (with 2 to 5 acres of land). The 1984 Campaign was thus to give special emphasis on reaching small farmers, and minimize the campaign "drop-outs" rate (farmers who discontinued practicing rat control).

4.1.1 Evaluation Methods

As shown in Figure 4-4, three types of evaluation studies were employed to assess the Campaigns' performance and effectiveness: (a) Management Monitoring Survey (MMS), (b) Information Recall & Impact Survey (IRIS), and (c) Field Damage Assessment Survey (FDAS).

The purpose of a Management Monitoring Survey (MMS) is to spot check whether a campaign strategy plan has been properly followed during implementation. It also investigates if the required inputs to support campaign activities have been made available and delivered as planned. MMS is useful in identifying management or implementation problems, and in analyzing alternative solutions or strategy changes which might be needed to improve campaign implementation before an activity is completed. In addition, MMS findings can help in validating the conclusion of an impact evaluation or information recall survey. For instance, in the 1983 Campaign, the IRIS found that comics were rarely cited by survey respondents as one of the sources of rat control information. However, MMS findings indicated that there was a problem in comics distribution. Therefore, it could be argued that the failure of comics as a campaign information source was primarily due to a management problem, and not necessarily due to a problem of message design or medium-selection strategy.

The Field Damage Assessment Survey (FDAS) is a means to assess the degree of damage caused by rats in a given area, and to determine the type/method used by farmers to control the extent of such a damage. The FDAS results, together with answers from farmers' interviews regarding the type of control method utilized, permit a comparison to be made of the relative efficiency of different control methods. Information on field damage reduction can be used to estimate the economic savings due to improved efficiency of rat control applications, and to measure the costs and benefits of campaign activities.

F I G U R E   4 - 4

Evaluation Procedures of 1983 and 1984 Campaigns

Year	Type of evaluation	Timing (after campaign started)	Survey coverage	Location
1983	1. Management Monitoring Survey (MMS)	2 to 3 weeks	10 district officers 29 upazila officers 43 union officers 305 farmers	10 primary districts only
	2. Field Damage Assessment Survey (FDAS)	2 months	851 plots (266 acres)	7 primary and 3 secondary districts
	3. Information Recall and Impact Survey (IRIS)	2 to 3 months	1,149 wheat farmers	6 primary (775 farmers) and 3 secondary (374 farmers) districts
1984	1. Management Monitoring Survey (MMS)	2 to 3 weeks	12 district officers 31 upazila officers 62 union officers 94 block supervisors	12 districts
	2. Field Damage Assessment Survey (FDAS)	2 months	866 plots (269 acres)	10 districts
	3. Information Recall and Impact Survey (IRIS)	2 to 3 months	1,089 farmers	9 districts

The Information Recall & Impact Survey (IRIS) is one of the most common evaluation tools used to examine the effectiveness of an extension or public education intervention programme such as a campaign. One of its purposes is to assess how campaign messages have been perceived, understood or accepted by target beneficiaries. Another important purpose is to determine whether target beneficiaries' levels of knowledge, attitude and practice regarding the campaign suggestions have changed as compared to their pre-campaign levels.

4.1.2 Evaluation Results

As can be seen from Figures 4-5a to 4-5d, evaluation findings indicated that the 1983 Campaign was successful in increasing the proportion of wheat farmers who conducted rat control. About 32 percent of wheat farmers in the survey reported to have practised rat control compared with only 10 percent before the campaign commenced (Fig. 4-5a). Although the 1983 Campaign was focused on rat control for wheat farmers, a limited spill-over effect, was also noted among non-wheat farmers. Such a marginal increase from 45 percent to 49 percent among all farmers (Fig. 4-5b), however, was not surprising, as the 1983 campaign was mainly designed to benefit wheat farmers.

After the 1984 Campaign, 67 percent of all farmers surveyed (including wheat farmers) reported to have practised rat control as compared with 49 percent before the campaign (Fig. 4-5b). Among wheat farmers, rat control practice increased from 32 percent in 1983 to 40 percent in 1984 (Fig. 4-5a). Since the 1984 Campaign was aimed at all types of farmers, a more substantial increase in rat control practice was noted among those farmers as compared to the increase among wheat farmers only. Such a case was expected, since most wheat farmers might already have been practising rat control as a result of the 1983 Campaign. Moreover, the 1984 Campaign was not specifically aimed at wheat farmers, but at all farmers. It was thus not surprising that the 1984 Campaign effects were more noticeable among all farmers as a group than just among the wheat farmers.

During the 1983 Campaign, the 11 wheat-growing districts (referred to as primary districts) were given more-intensive campaign treatment (i.e., information exposure) as compared with the remaining 10 non-wheat-growing districts (or secondary districts). As expected, the 1983 Campaign was more effective in increasing rat control practice among farmers in primary districts from 9 percent before the campaign to 33 percent after the campaign (Fig. 4-5c). In the secondary districts, it only increased from 14 percent to 29 percent. The data also showed that rat control prior to the 1983 Campaign was practised more in non-wheat-growing districts than in wheat-growing districts. The 1983 Campaign strategy of positioning campaign messages to wheat farmers as the primary target beneficiaries was thus appropriate and successful in increasing rat control practice among these farmers.

As shown in Figure 4-5d, the 1983 Campaign was more successful in increasing rat control practice among large farmers (with more than 5 acres of land), and to a small extent, among medium farmers (with 2-5 acres of land). No increase was noted among small farmers (with less than 2 acres). The 1983 Campaign appeared to have widened the rat control practice gap between small farmers and large or medium farmers which was only 6 percent before the campaign. After the campaign, the gap widened to 17 percent between the small and the large farmers, and to 8 percent between the small and the medium farmers. Likewise, the gap also widened between medium and large farmers, from 0 percent to 9 percent (Fig.4-5d).

In recognition of such problems, the 1984 Campaign devised additional strategies for narrowing the campaign effects gap between small farmers and large or medium farmers. Among others, message "redundancy" and/or "ceiling-effects" strategies were used. These strategies seemed to have worked quite satisfactorily, as indicated in the results of the 1984 Campaign. For instance, rat control practice among small farmers was reported to have increased considerably, from 41 percent to 63 percent after the 1984 Campaign. Among the large farmers, the increase (14 percent) was significantly less than that of the small farmers (22 percent).

FIGURE 4-5

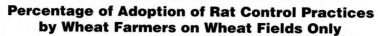

Summary of Main Evaluation Findings of the 1983 and

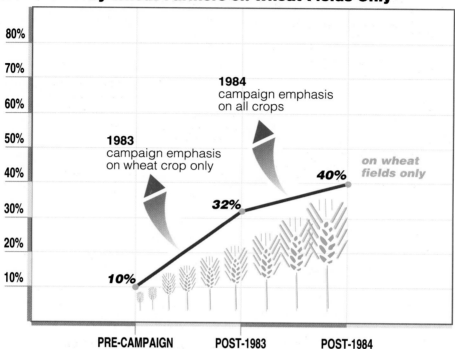

a

Percentage of Adoption of Rat Control Practices by Wheat Farmers on Wheat Fields Only

1984
campaign emphasis
on all crops

1983
campaign emphasis
on wheat crop only

on wheat fields only

40%

32%

10%

PRE-CAMPAIGN   POST-1983   POST-1984

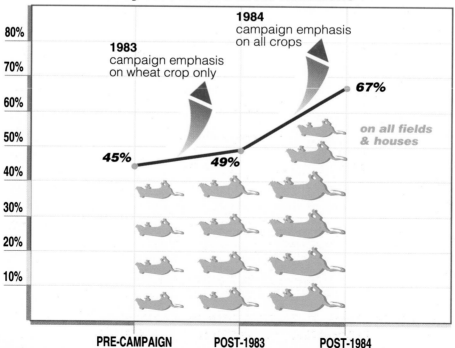

b

Percentage of Adoption of Rat Control Practices by All Farmers in All Locations

1984
campaign emphasis
on all crops

1983
campaign emphasis
on wheat crop only

67%

on all fields & houses

45%   **49%**

PRE-CAMPAIGN   POST-1983   POST-1984

c

Percentage of Adoption of Rat Control Practices by Wheat Farmers on Wheat Fields Primary and Secondary Districts

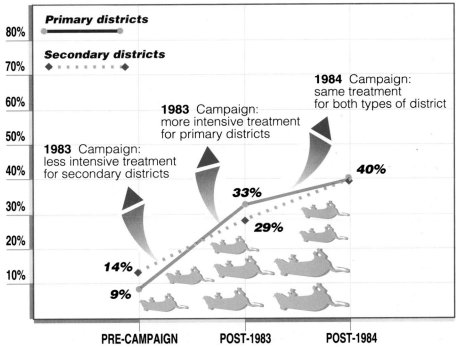

1984 Campaign: same treatment for both types of district

1983 Campaign: more intensive treatment for primary districts

1983 Campaign: less intensive treatment for secondary districts

40%

33%

29%

14%

9%

PRE-CAMPAIGN   POST-1983   POST-1984

d

Percentage of Adoption of Rat Control Practices by Large, Medium and Small Farmers in All Locations

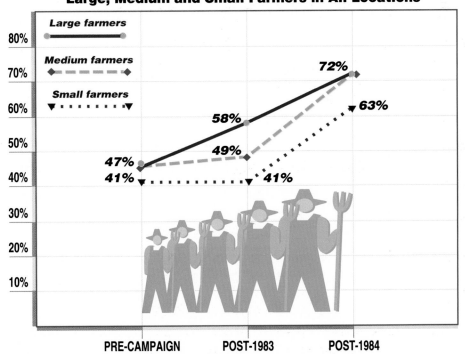

72%

63%

58%

49%

47%

41%   41%

PRE-CAMPAIGN   POST-1983   POST-1984

The 1984 Campaign was able to increase the number of farmers who practised rat control and to reduce the widening gap created during the 1983 Campaign between large and small farmers in terms of rat control practice. Nevertheless, such a gap after the 1984 Campaign (9 percent) was still wider than before the 1983 Campaign (6 percent). This gap may have been due not only to socio-economic disadvantages accruing to small farmers, but perhaps also because the adoption of rat control practices depends on the extent of rat problems they faced. Farmers who own or cultivate large parcels of land are likely to have more rat problems than do small farmers. Thus the need for rat control is felt more keenly among large farmers than small farmers.

4.1.3   Costs and Benefits of the Campaigns

The 1983 Campaign required about four months of planning and preparation, and about US$ 17,000 in operational expenses (Fig. 4-3b). The largest expenditure was for developing and printing the campaign media materials, which accounted for 59 percent of the total campaign operational costs. Salaries of the campaign planners and staff were not included in the campaign cost, since such expenditures would have been incurred anyway even if the campaign had not been held. In other words, no additional manpower resources were mobilized by the DOAE for the campaign.

According to the evaluation findings, 43 percent of the survey respondents were exposed to the 1983 Campaign. It was estimated that about 4.6 million farm households were "reached" by the campaign. The average cost of "campaign-reach" per farm household was thus less than US$ 0.01 which was very low by any campaign standard. It was recognized, however, that not all farmers who had been exposed to the campaign messages would automatically conduct rat control. Although the cost of informing, motivating, and educating farmers to conduct rat control could be estimated, it would probably be more useful to calculate the benefits gained from the rat control campaign.

The Field Damage Assessment Survey (FDAS) of the 1983 Campaign found the damage level of wheat fields treated with ready-made baits, as recommended by the campaign, was on average 56 percent lower than those not given any treatment (Fig. 4-6a). Using wheat as an indicator crop, the FDAS estimated that the wheat production gain due to ready-made bait application alone was 5,045 tons, with a market value of US$ 834,000 (Fig. 4-6b), in one winter crop season. Since the cost of ready-made bait treatment was US$ 23,000, the net gain during the 1983 wheat season was approximately US$ 834,000. This also means that a wheat farmer who practised rat control with ready-made bait saved about US$ 8.90 per hectare of land. It should be pointed out that this benefit analysis was limited to the application of ready-made baits on wheat fields only. The campaign, as discussed earlier, was also successful in increasing the number of farmers who conducted rat control in non-wheat fields, and by using other rat control methods. However, no specific study was conducted to calculate such cost-benefits.

With the benefits of experience and skills in planning and developing extension campaign strategies and the multi-media support materials gained during the 1983 Campaign, it only required about 2 months to prepare the 1984 Campaign. The campaign operational cost was about US$ 41,000 as it covered all 21 districts in the country. The IRIS indicated that an estimated of 67 percent of respondents surveyed had been exposed to 1984 campaign messages. It was thus estimated that about 7.2 million farm households had known about the campaign. This might have also been the result of the cumulative effects of both the 1983 and 1984 Campaigns.

The 1984 FDAS again used wheat as the indicator crop to estimate the benefits owing to ready-made bait application. As shown in Figure 4-7a, when damage levels between untreated wheat fields and those treated with ready-made baits were compared, a significant damage reduction of 41 percent was noted. It was estimated that about 5,208 tons of wheat with a market value of US$ 859,000 was saved from rats by applying ready-made baits alone, in one winter-crop season, as a results of the 1984 Campaign. Since the cost of treatment was about US$ 37,000, the net gain due to the use of ready-made baits in wheat fields was approximately US$ 822,000 as shown in Figure 4-7b. As the 1984 Campaign was also successful in increasing the proportion of farmers who conducted rat control in non-

Comparison of Rat Control Methods Applied in Wheat Fields during or after the 1983 Campaign

Control method	Numbers of plots	Percentage of damage	Percentage of damage reduction relative to no control	T-test
No control	209[1]	4.44	–	–
Rodenticides other than ready-made bait	57	3.27	26%	not significant
Ready-made bait	109	1.97	56%	p<0.005

Note:
1) Since farmers control rats only in infested or damaged fields, fields with zero damage were not included in the group with no control

Economic Benefits from the Application of Ready-Made Bait in Standing Wheat after the 1983 Campaign

Wheat area treated with ready-made baits [1] (18% x 519,000 ha)	93,420	ha
Yields in infested wheat fields [2] :		
Treated fields (potential yield less damage [3]) : 2,157 kg/ha - 1.97% x 2,157 kg/ha	2,115	kg/ha
Untreated fields (potential yield less damage [4]) : 2,157 kg/ha - 4.44% x 2,157 kg/ha	2,061	kg/ha
Difference due to ready-made bait treatment	**54**	**kg/ha**
Production gain due to treatment: (area treated x difference) = 93,420 ha x 54 kg/ha	**5,045**	**tons**
Price per ton (c.i.f. Chittagong) = US $ 170		
Value of production gain = 5,045 x $ 170	US $	857,650
Cost of treatment [5]	US $	23,355
Net gain due to ready-made bait treatment during the 1983 wheat season	**US $**	**834,295**

Note:
1) Based on the 1983 field damage assessment survey (18% of the total wheat-producing areas were treated with ready-made baits) and *FAO production yearbook,* vol. 37, 1983 (total wheat-producing area in Bangladesh was 519,000 ha).
2) The *FAO production yearbook,* vol. 37, 1983 estimated for Bangladesh, an average wheat yield of 2,109 kg/ha which included 2.21% average rat damage.

Thus, potential wheat yield in 1983 was: $\dfrac{2,109 \text{ kg/ha} \times 100}{100 - 2.21} = 2,157$ kg/ha

3) Percentage of damage on fields treated with ready-made baits was 1.97% (Figure 4-6a)
4) Percentage of damage on untreated fields was 4.44% (Figure 4-6a)
5) Based on 5 Taka per hectare, excluding labour (about 1 hour) and US $ 1 = 20 Taka (exchange rate in 1983)

Costs and Benefits of the Campaigns

Comparison Between Rat Control Methods Applied in Wheat Fields during or after the 1984 Campaign

Control method	Numbers of plots	Percentage of damage	Percentage of damage reduction relative to no control	T-test
No control	165[1]	4.90	–	–
Rodenticides other than ready-made bait	130	4.65	5%	not significant
Ready-made bait	130	2.88	41%	p<0.001

Note: The difference in damage inflicted on fields treated with rodenticides other than ready-made baits and with the ready-made baits was significant at the 0.01 level.
  1) Since farmers control rats only in infested or damaged fields, fields with zero damage were not included in the group with no control

Economic Benefits from the Application of Ready-Made Bait in Standing Wheat after the 1984 Campaign

Wheat area treated with ready-made baits[1] (19% x 623,000 ha)	118,370	ha
Yields in infested wheat fields[2] :		
Treated fields (potential yield less damage[3]) : 2,162 kg/ha - 2.88% x 2,162 kg/ha	2,100	kg/ha
Untreated fields (potential yield less damage[4]) : 2,162 kg/ha - 4.90% x 2,162 kg/ha	2,056	kg/ha
Difference due to ready-made bait treatment	**44**	**kg/ha**
Production gain due to treatment: (area treated x difference) = 118,370 ha x 44 kg/ha	**5,208**	**tons**
Price per ton (c.i.f. Chittagong) = US $ 165		
Value of production gain = 5,208 tons x US $ 165	US $	859,320
Cost of treatment[5]	US $	36,991
Net gain due to ready-made bait treatment during the 1984 wheat season	**US $**	**822,239**

Note:
1) The 1984 field damage assessment survey found that 19% of the total wheat-producing area was treated with ready-made baits, and on the basis of the 1983 wheat-producing area of 519,000 ha (*FAO production yearbook,* vol. 37, 1983) the 1984 total wheat-producing area in Bangladesh was estimated at 623,000 ha.
2) The *FAO production yearbook,* vol. 37, 1983 estimated wheat yield of 2,109 kg/ha for Bangladesh with a 2.45% average rat damage.

Thus potential wheat yield in 1984 was about $\dfrac{2,109 \text{ kg/ha} \times 100}{100 - 2.45} = 2,162 \text{ kg/ha}$

3) Percentage of damage on fields treated with ready-made baits was 2.88% (Figure 4-7a)
4) Percentage of damage on untreated fields was 4.90% (Figure 4-7a)
5) Based on 6.24 Taka per hectare, excluding labour (about 1 hour) and US $ 1 = 20 Taka (exchange rate in 1984)

wheat fields, and by using other control methods, the total benefits were much greater than the figure calculated for wheat alone.

The most significant result of these campaigns, however, was the increased support and recognition among Bangladesh's agricultural officials and extension staff on the need for a systematic and strategically-planned extension programme. The campaigns have empirically demonstrated the importance and strategic role/function of a well-planned and integrated extension activity utilizing a multi-media campaign approach. Having been involved directly in such an activity, and having seen positive results of the campaigns, many researchers, subject-matter specialists, extension officers, trainers, and communication support personnel, are now likely to work more closely as a team, especially at the extension programme planning stage, and in field-level implementation.

In addition, the empirical campaign evaluation results have convinced several major local rodenticide manufacturers that a market demand exists for good quality ready-made baits and that a regular and mass production of such baits is economically feasible. The effectiveness of a rat control programme can be threatened in the absence of a continuous and easily accessible supply of reasonably priced and good quality rodenticides, which was the case prior to the 1983 Campaign.

4.2.
Malaysia:
the Rat Control Campaign

Starting in early 1985, Malaysia's Department of Agriculture (DOA) has embarked on a systematic extension programme which integrated technology generation, extension, training and multi-media materials development activities into a coherent and well-planned campaign. The first of such an undertaking was carried out by DOA in collaboration with FAO's Inter-Country Programme on Integrated Pest Management (IPM) in Rice in South and Southeast Asia, with technical assistance from FAO's Agricultural Education and Extension Service (ESHE). Due to severe rat problems in Penang, Malaysia, a Strategic Extension Campaign on Rat Control was conducted following the SEC process and methodology applied by FAO earlier in Bangladesh. However, having learned from the lessons and experiences of Bangladesh, two main elements were added in the Malaysia's SEC replication :

a   a comprehensive Knowledge, Attitude and Practice (KAP) Survey, which utilized both sample survey (for quantitative data) and focus group interview/discussion (for qualitative info.) methods.

b   a series of SEC skills-oriented workshops (whose topics are listed in Fig. 2-3) for a core-group of about 25-30 future SEC planners & trainers, consisting of subject-matter specialists, extension planners/managers, trainers, field workers, communication support staff, etc. (hereafter referred to as the "Core-Group")

The Malaysian Rat Control Campaign activities followed closely the SEC Process and Implementation Steps outlined in Fig. 2-2. It was the first complete and comprehensive SEC programme which had been conducted in Malaysia, and the first SEC programme supported by FAO which carried out the entire SEC process and had all the suggested SEC elements.

The SEC Process

→ KAP survey

The SEC activity started with a one-week workshop for the Core-Group on the purpose and methods of Knowledge, Attitude, and Practice (KAP) Survey. During this workshop, 30 participants together with some farmer leaders were requested identify information or data which would be critical, and thus need to be obtained, for formulating the campaign objectives and developing the strategic extension campaign plan. Assisted by, and working together with, the KAP Survey investigators/researchers, the Core-Group then developed the data collection instrument for the KAP Survey. Such an approach in developing action-oriented survey questions can help ensure the relevance and practicality of the survey results for SEC planning, management and evaluation purposes, rather than just for an academic exercise. The field implementation (e.g., data collection, processing, analysis, and reporting) of the KAP Survey itself was contracted to the Agricultural University of Malaysia's Centre for Extension and Continuing Education, coordinated by Dr. Azimi Hamzah.

KAP Workshop in session

KAP Survey and Focus Group Interviews

→   **Problems Identification**

→   **Objectives Formulation**

Upon completion of the KAP Survey which took about 2.5 months, a two-week workshop for the same Core-Group was held to train them on the concepts, methods and skills of campaign strategy development and planning, message design and prototype multi-media materials development. The survey results were the main inputs used by the Core-Group to formulate the SEC objectives. For example, problems related to inappropriate or non-adoption of rat control were identified based on the findings of the KAP Survey of farmers in Penang State. Then, a set of specific, measurable, and problem-solving SEC objectives were formulated by the Core-Group during this workshop, in consultations with some farmers' representatives. To illustrate the degree of specificity of the campaign, the list of problems is shown in Fig. 4-8, and the campaign objectives can be seen in Fig. 4-9.

F I G U R E   4 - 8

Identified Problems in Rat Control in Penang, Malaysia
Based on findings of a KAP survey of farmers in Penang State

1. Low knowledge of the value of physical methods and cultural practices regarding rat control

2. Low knowledge of different functions and characteristics of different rodenticides

3. Misconception that rats are "intelligent", and thus unlikely to be successfully controlled

4. Lack of group and collaborative efforts to control rats

5. No action to control rats until damages are visible

6. Inappropriate application of rodenticides in different situations

7. Most farmers have more than one job and thus do not have enough time to control rats

8. Superstition that rats would take revenge on behalf of their dead "friends" by causing worse damages

9. Simultaneous planting is not practised, thus providing continuous food supply for rats

---

Source: R. Adhikarya (1985), "Planning and Development of Rat Control Campaign Objectives
and Strategies for the State of Penang, Malaysia".
Note: KAP refers to knowledge, attitudes and practice of the target audience.

Figure 4-9

Specific and Measurable Campaign Objectives for Rat Control Campaign in Penang State, Malaysia

Identified Problems among Farmers	Formulated Extension Campaign Objectives (based on KAP survey results)
1. Inadequate knowledge of the value of physical methods and cultural practices regarding rat control	To raise the proportion of rice farmers' level of knowledge/appreciation concerning the value and benefits of cultural practices from 67% to 75%, and physical rat control practices from 31% to 45%
2. Little knowlege of the different functions and characteristics of different rodenticides	To raise the proportion of rice farmers' level of awareness and knowledge by improving their understanding regarding the different functions and characteristics of two types of rodenticides: a) chronic poison baits from 61% to 70% b) chronic poison dust from 22% to 40%
3. Misconception that rats are "intelligent" and thus unlikely to be successfully controlled	To reduce the proportion of rice farmers' misconception that rats areunlikely to be controlled successfully because they are "intelligent" from 52% to 35%
4. Lack of group and collaborative efforts in controlling rats	To encourage greater participation of rice farmers in group and/or collaborative efforts in controlling rats, by increasing the proportion of rice farmers' level of favourable attitudes towards such efforts from 60% to 70%
5. Farmers normally do not take voluntary action to control rats until crop damages are visible	To increase the proportion of rice farmers who believe that rat control is not a waste time from 55% to 65% in order to encourage them to take action before their crops are damaged
6. Inapproppriate application of different rodenticides in different situations/stages	a) To increase the proportion of rice farmers' knowledge on the correct application of rodenticides with regard to: 1. Rate of application of acute poison from 11% to 40 % ; chronic poison (baits) from 23% to 40%; and chronic poison (dust) from 67% to 75% 2. Time of application of acute poison from 47% to 60%; chronic poison (baits) from 39% to 50%; and chronic poison (dust) from 41% to 55% 3. Location to place acute poison from 43% to 55%; chronic poison (baits) from 43% to 55%; and chronic poison (dust) from 78% to 80% b) To increase the proportion of rice farmers' level of appropriate practice in rodenticides application with regard to: 1. Rate of application of acute poison from 12% to 24%; chronic poison (baits) from 23% to 40%; chronic poison (dust) from 32% to 40% 2. Time of application of acute poison from 28% to 35%; chronic poison (baits) from 28% to 35%; chronic poison (dust) from 43% to 50%
7. Lack of motivation of most farmers who have more than one job, to spend more time and effort to control rats in order to increase their yields and income	To motivate and encourage rice farmers to spend more time and efforts to control rats in order to increase their crop yields and income, by increasing their perception that controlling rats is more beneficial than doing other jobs; from 37% to 50%
8. Superstition that rats will take revenge on behalf of their "dead friends" by causing worse damages	To reduce the proportion of rice farmers' misconception regarding their superstitious belief that rats take revenge on behalf of their "dead friends" by causing worse damages from 54% to 50%
9. Non-practice of simultaneous planting which could disrupt food supply for rats during part of the year	To encourage more rice farmers to engage in simultaneous planting in order to reduce time for rats to have continuous food supply by enhancing positive attitudes towards that practice; from 79% to 85%

Source: Adapted from R. Adhikarya (1985)

→ **Strategy Development & Planning**

→ **Message Design & Positioning**

→ **Prototype Multi-Media Materials Development & Packaging**

The campaign was targeted to about 14,000 farm families in 27 Extension Service Areas and 1,197 Contact Farmers in Penang State. To reach these target groups effectively, a detailed campaign strategy plan and prototype message design and multi-media materials were also developed by the Core-Group during this two-week workshop. Examples of the strategy development and message design/positioning worksheet are presented in Fig. 4-10a and Fig. 4-10b.

Campaign Planning and Strategy Development Workshop

Campaign Planning and Multi-Media Materials Development

Example of a Worksheet for:
Extension Planning and
Strategy Development Process Exercise

Problem	Reasons or causes for problem	Problem-solving strategy or approach	Information positioning approach
Farmers' misconception that rats take revenge on behalf of their "dead friends" by causing worse damages	Superstitious belief	As a Moslem, it is sinful to believe in superstitions. Other specific citation from the Holy Book (Quran or Hadith) regarding the above	Religious disincentive
Not enough time to control rats	70% of farmers have more than one job	"Rat control method of using wax and dust poison is simple and easy. Even your wife and children can do it if you are busy".	Task delegation
Inappropriate application of rodenticides at different crop growing stages	Unclear and complicated rodenticide application recommendations	Simplification of technology recommendations and easy-to-remember application procedures	Simplicity
	Inefficent use (too strong a dosage) of rodenticides since they are distributed free to farmers	Arouse guilt feeling of farmers by stressing the waste of their community funds due to inefficient and ineffective use of the free rodenticides	Guilt feeling creation
Farmers' misconception that rats are "intelligent", thus control unlikely to succeed	Failure of zinc phosphide (e.g. bait shyness effect) which is used by the majority of farmers	De-emphasize the use of acute poison (zinc phosphide) and encourage the use of wax poison before booting stage and dust poison after booting stage.	Down-playing the competitor. Easy-to-remember action.
		Stress the need for group and collaborative efforts instead of the individual approach if the battle to fight "smart" rats is to be won	Need for group efforts

Source: Adapted from R. Adhikarya (1985)

FIGURE 4-10 b

Example of a Worksheet for:
Message Design Process Exercise

Problem solving strategy	Message appeals	Examples of message appeals	Channel of message-delivery
Counter-attack farmers' superstitious belief that rats take revenge on behalf of "dead friends" by causing worse damages	Fear arousal	"Its is sinful for a Moslem to believe in superstition"	Religious leaders' sermons during Friday prayers; leaflet; radio spots
	Religious Incentive	"The more rats you kill, the more you will be rewarded in heaven" (Citation from the Holy Book of Islam)	
Discourage the use of zinc phosphide and encourage the wax and dust poison (chronic rodenticides)	Safety convenient/ simplicity	"Wax poison and dust poison are much safer, easier and more effective than zinc phosphide"	Instructional poster; radio spots; pamphlet; portable flipchart; pictorial booklet
	Testimonials	Use testimony from safisfied wax and dust poison adopters about its simplicity of use and effectiveness	
Motivating farmers to conduct group/ collaborative action in controlling rats (e.g. simultaneous planting, applying physical control methods, or conducting simultaneous rat control)	Ridicule	"Since rats are 'intelligent' if you fight them alone you might lose and thus be a victim", "let's do it together"	Motivational poster; leaflet; pictorial booklet; slide-sound
	Cultural/ traditional value	"Gotong-Royong (working together and helping each other in a community) is a virtue and the most effective means to control rats"	
	Solidarity	"Bersatu kita teguh, bercerai kita roboh" (united we stand, divided we fall)	
Educating farmers on the appropriate use/application of rodenticides	Guilt feeling, civic responsibility	"Don't you feel guilty wasting community funds if you are not using the free rodenticides properly"	Instructional poster; radio spots; group discussions; portable flipchart
	Easy to remember	"Use wax poison weekly before booting stage, and dust poison after booting stage"	

Source: Adapted from R. Adhikarya (1985)

→   **Pretesting of Prototype Multi-Media Materials**
→   **Revision & Mass Production of Materials**

The prototype multi-media materials produced during the above mentioned workshop were then pretested by the Core-Group in the field with a sample of the actual target audience during another one-week workshop specially designed for this purpose. Based on the findings and suggestions of the field pretesting, the prototype materials were then revised and mass-produced accordingly.

Field Pretesting of Campaign Materials.

Field Pretesting of Campaign Materials.

Some of the Campaign Multi-Media Materials.

→ **Management Planning**
→ **Briefing & Training of Field Personnel**
→ **Distribution of Multi-Media Materials**
→ **MMS & IRIS Preparations**

In another one-week workshop, the Core-Group developed a Management Plan which detailed the procedures for the campaign implementation, including the specific tasks and responsibilities of extension personnel at the field level. The distribution and utilization plan for the multi-media support extension and training activities was also developed. Special briefing and training sessions for field personnel involved in the SEC activities were held to ensure they understood their role and functions, and properly utilized the multi-media materials to support their extension and training activities. Most of these materials were distributed to the appropriate users during the SEC briefing and training sessions. On the basis of this Management Plan, the design and implementation plan for the Management Monitoring Survey (MMS) were made. The Core-Group also consulted with the survey investigators/researchers on the procedures and information/data which had to be collected for the MMS and the Information Recall and Impact Survey (IRIS), during a one-week workshop.

Campaign Management Planning Session

Field Personnel Training and Distribution of Materials

→ **Field Implementation**

The Campaign was officially launched on 11 October 1986 with an elaborate Opening Ceremony attended by important National and State officials. It was given wide publicity coverage by the national and local press, radio and television. The duration of the campaign was for 4-5 months, during which time the various activities following the strategy plan were undertaken in the rice-growing districts of Penang State.

Campaign Launching Ceremony and Field Extension Workers Singing Rat Control Songs

*Extension Workers Motivating and
Educating Farmers on Rat Control*

Practicing Rat Control in the Field

→   **Evaluation Studies :** *Management Monitoring Survey*
Information Recall & Impact Survey

Two weeks after launching the campaign, a Management Monitoring Survey (MMS) was conducted and it was completed within less than 4 weeks by Dr. Ramli Mohamed and Dr. Yoke-Lim Khor of the Science University of Malaysia. Thus by late November 1986, the campaign planners and managers got the initial feedback as to how the campaign was going, and what management or logistical problems were encountered in the field during the first four weeks of campaign implementation. By the beginning of the second month of the campaign period, necessary changes and modifications of its strategy and management plans were made accordingly.

The Information Recall & Impact Survey (IRIS) was basically a "Post-Campaign" KAP Survey. It was conducted by the same survey investigators from the Agricultural University of Malaysia who did the "Pre-Campaign" KAP Survey, in order to facilitate the comparative data analysis of the "before" and "after" campaign results. The IRIS was conducted in April 1987, — about 6 weeks after the official completion of the campaign period. A Focus Group Interview/Discussion (FGI) was also carried out to complement IRIS. Figure 4-11 shows the information on the number of respondents of the various surveys conducted for the SEC on Rat Control in Malaysia.

Conducting Management Monitoring Survey and Information Recall & Impact Survey

FIGURE 4-11

Evaluation Studies for Malaysia's Rat Control Campaign			
Type of survey	Time period	Duration	Sample/respondents
KAP Survey (pre-campaign)	Jun. - Aug. 1985	2.5 months	200 rice farmers
Management Monitoring Survey	Nov. 1986	1 month	213 rice farmers 61 contact farmers 17 agricultural officers/extension workers 15 religious/local leaders
Information Recall & Impact Survey (post-campaign)	Apr. - Jul. 1987	3 months	415 rice farmers
Focus Group Interviews/discussions	Aug. 1987	1 month	9 farmers' groups (with average of 8 farmers each)

4.2.2   *Evaluation of SEC Results*

Some of the important evaluation findings of the campaign are presented in the following pages in a graphical manner. These evaluation results showed the changes in terms of the knowledge, attitude and practice levels of rice farmers in Penang vis-a-vis rat control campaign recommendations and messages. Almost all of the targets of the campaign objectives had been accomplished. As a result of the campaign, the number of farmers who reported experiencing damages caused by rats had also declined. For instance, there was a significant decrease in the proportion of farmers who reported that all the rice plant damages were due to rats from 47 percent before the campaign to 28 percent after the campaign. The Malaysian Department of Agriculture (DOA) data indicated that in Penang the rice field damages due to rats in 1984 (before the campaign) was about 700 ha. compared to only 223 ha. in 1988 (Mohamed, 1989).

Rice Farmers' KNOWLEDGE of Rat Control Measures

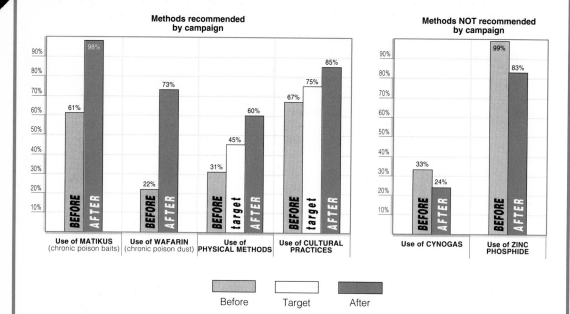

Methods recommended by campaign

Methods NOT recommended by campaign

Before — Target — After

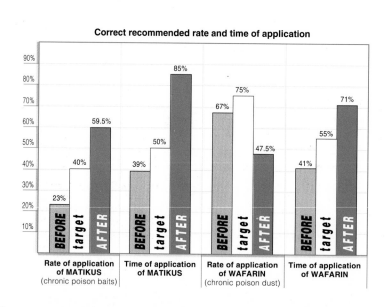

Correct recommended rate and time of application

Sources of the Figure above and those in pages 121-123:

For data/info. on "Before Campaign" (Pre-campaign KAP Survey findings): A. Hamzah & J. Hassan, "Rice farmers' knowledge, attitude and practice of rat control: a study conducted in Penang, Malaysia", Serdang: Agricultural University of Malaysia (August, 1985).

For data/info. on "After Campaign" (Post-campaign KAP Survey findings): A. Hamzah & E. Tamam, "Rat control strategic multi-media campaign for the state of Penang, Malaysia. A report of the information recall and impact survey", Serdang: Agricultural University of Malaysia (August, 1987).

Evaluation of SEC Results

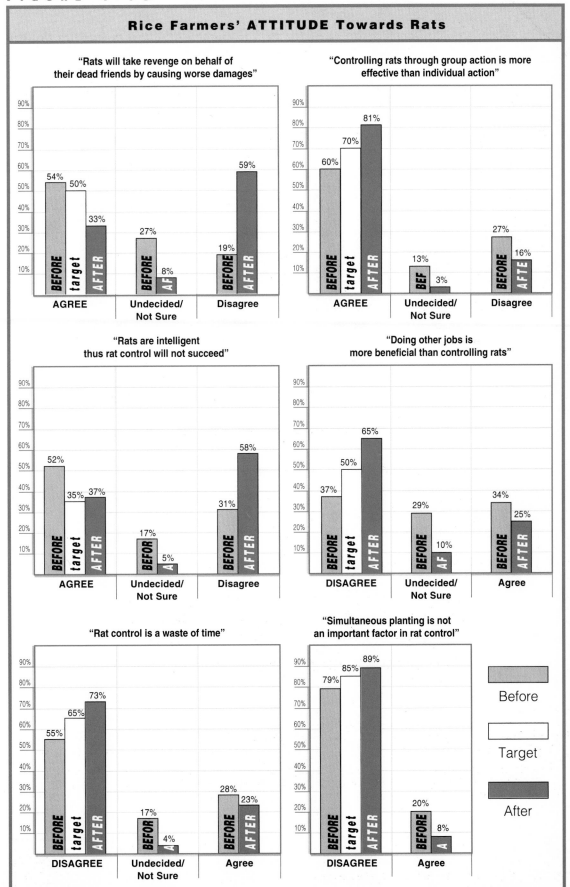

Rice Farmers' ATTITUDE Towards Rats

"Rats will take revenge on behalf of their dead friends by causing worse damages"

"Controlling rats through group action is more effective than individual action"

"Rats are intelligent thus rat control will not succeed"

"Doing other jobs is more beneficial than controlling rats"

"Rat control is a waste of time"

"Simultaneous planting is not an important factor in rat control"

Before

Target

After

Sources: A. Hamzah and J. Hassan (1985) for "Before Campaign" data/information; A. Hamzah and E. Tamam (1987) for "After Campaign" data/information.

FIGURE 4-14

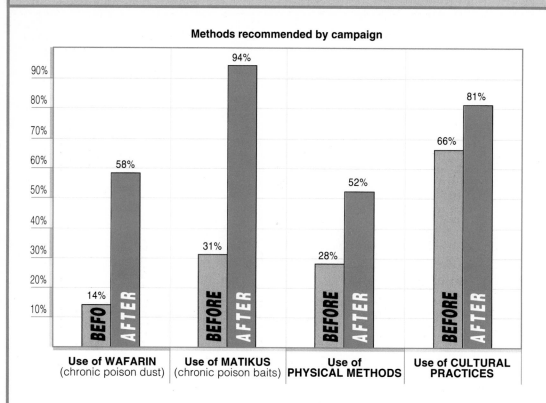

Rice Farmers' PRACTICE of Rat Control Methods

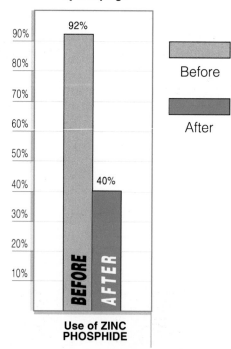

Sources: A. Hamzah and J. Hassan (1985) for "Before Campaign" data/information;
A. Hamzah and E. Tamam (1987) for "After Campaign" data/information.

FIGURE 4-15

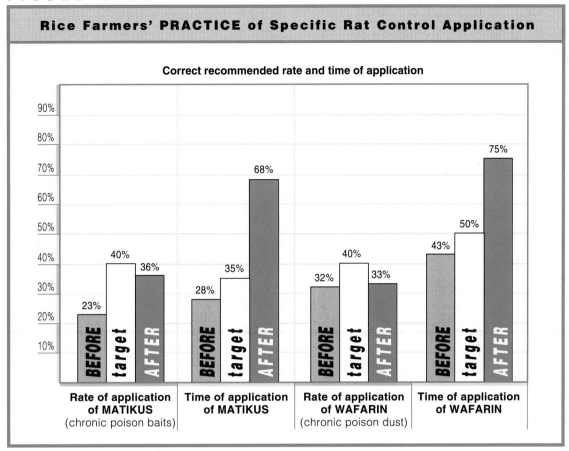

Rice Farmers' PRACTICE of Specific Rat Control Application

Correct recommended rate and time of application

Sources: A. Hamzah and J. Hassan (1985) for "Before Campaign" data/information;
A. Hamzah and E. Tamam (1987) for After Campaign" data/information.

FIGURE 4-16

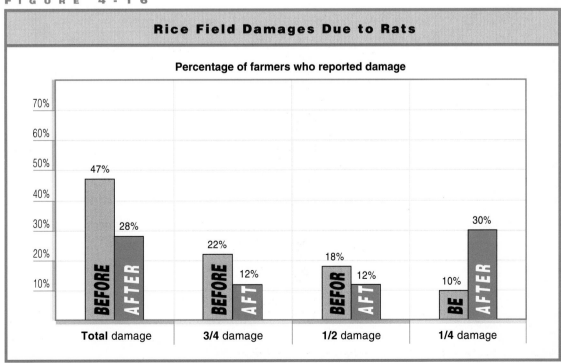

Rice Field Damages Due to Rats

Percentage of farmers who reported damage

Sources: A. Hamzah and J. Hassan (1985) for "Before Campaign" data/information;
A. Hamzah and E. Tamam (1987) for "After Campaign" data/information.

Costs and Benefits of SEC

The data provided by the DOA showed that about 477 ha. or 1,178 acres of rice fields in Penang were saved from rat damage due to the campaign (Mohamed, 1989). As shown in Fig. 4-17, it was estimated that on the basis of 1,600 kg. rice yield per 1 acre, about 1,885 tons of rice were saved in one season alone after the campaign. Therefore, a saving of about US$ 365,180.- was made (based on a market value of US$ 193.75 per ton of rice). As the campaign was aimed at about 14,000 farm families, the economic benefit to the community can be estimated at an average of US$ 26.- per farm family, during one season.

The Information Recall & Impact Survey (IRIS) data indicated that about 34 percent of all farmers surveyed were practising rat control at a high degree of effectiveness/appropriateness (Hamzah and Tamam, 1987). Although other farmers reported to have conducted rat control, they might not have applied all the correct or recommended methods or procedures. Since obviously not all the 14,000 farm families were reached, adopted, or properly applied the recommended rat control methods, the economic benefit to those who did effectively control rats in their rice fields was much higher than the average estimate of US$ 26.- per farm family.

There are several reasons which made the campaign expenditure which totalled about US$ 140,184 extraordinarily high. First, it should be noted that the cost included training the Core-Group of 34 persons in five workshops for a total of 32 working days. As can be seen in Fig. 4-18, this training expenditure accounted for about 12 percent of the total campaign cost. Such a training was necessary as they had never been trained on the SEC process and methods before. Thus, this is a non-recurrent human resources development investment cost. Second, although the campaign only covered a relatively small area and was targeted to a limited number of farm families, the cost for evaluation studies could not have been proportionately reduced vis-a-vis the cost of conducting a large-scale survey for a larger or nation-wide campaign. The cost for these four evaluation studies was about US$ 22,547 or 16 percent of the total campaign expenses. Third, the input/supply expenditures, i.e. the rat baits, which accounted for about 9 percent, were also included in the campaign cost.

FIGURE 4-17

Cost and Benefit Analysis of Malaysia's Rat Control Campaign (in US $)

Acreage loss [1]:	1984 =	700 ha	or	1,729 acres
	1988 =	223 ha	or	551 acres
Production estimates:	1 acre =	1,600 kg [2] =		$310
Financial loss:	1984 =	2,766,400 kg =	$ 535,990	
	1988 =	881,600 kg =	$ 170,810	
Difference =		1,884,800 kg [2] =	$ 365,180	

Total savings = $ 365,180
Total expenditure for campaign = $ 140,184

Cost/benefit ratio   **1 : 2.60**

For each $ 1 invested, a return of $ 2.60 was gained.

Campaign Target: 14,000 farm families

Average Economic Benefit per Farm Family: $ 365,180 : 14,000 = $ 26

Note: 1) Reported by Mohamed (1989) based on data from Malaysia's Dept. of Agriculture.
2) Based on a market value of US $ 193.75 per ton of rice in 1988.

FIGURE 4-18

Expenditure for Malaysia's Rat Control Campaign (in US $)

	Items	Cost in US $	%
1	Training/Workshops	16,456	11.7
2	Multi-media materials development & production (including radio programmes)	72,989	52.1
3	Essay competition prizes	1,845	1.3
4	Rat baits	12,497	8.9
5	Campaign opening ceremony	922	0.7
6	Field implementation (field workers briefing/training)	12,928	9.2
7	Evaluation studies: a KAP $ 4,612 b. MMS $ 5,936 c. IRIS $ 5,904 d. FGI $ 6,095	22,547	16.1
Total		**140,184**	**100**

The cost for the multi-media materials development and production was quite high, about 52 percent of the total campaign expenditure. In terms of proportion, however, it was at about the same level as that of the Bangladesh's Rodent Control Campaigns (59 percent for 1983 and 49 percent for 1984). It is usually quite normal to expect a high proportion of a campaign cost for materials production (especially if technology inputs/supplies are not included in the cost analysis). However, in the case of Malaysia's Rat Control Campaign, it was unnecessary to produce printed campaign materials such as posters, flipcharts, pamphlets, pictorial booklets, etc. with high-quality, imported and glossy paper. Due to some "non-technical" considerations, DOA with its own resources (thus not using FAO project funds) opted to go "high-profile" in demonstrating the concrete products of its staff who, in less than 3 months, were able to design, develop, package, pretest, revise and produce these multi-media support materials for use in their field extension and training activities. Never before had they produced such well-researched and appropriately designed multi-media materials in a wide-ranging variety for various target groups and users, and in such a short time period. Hence, the opportunity to "show-case" such an excellent staff performance and the concrete SEC training outputs.

Even with a relatively high campaign cost, the cost-benefit ratio showed a favourable rate of return. As shown in Fig. 4-17, it was estimated that for each US$ 1.- spent on the campaign, a return of about US$ 2.60 was gained, and the economic benefit per family averaged about $ 26.- for the one season when the campaign was launched. It should be pointed out that Government of Malaysia's contribution to these SEC activities was much larger than that of the FAO project which provided the training costs, evaluation expenses, and technical assistance, amounting to probably less than 35 percent of the total SEC cost. This suggests the high-level policy and budgetary commitment and support of the Government to the SEC approach.

4.2.4. *Sustainability of SEC in Malaysia*

In a separate survey to assess the impact of the SEC training among the workshops' participants, it was found that many members of the Core-Group had applied the SEC knowledge and skills they acquired in their routine work programme (Mohamed, 1989). The following are some examples of the results of such SEC applications by the Core-Group members and/or others whom they had trained :

→ In the Selangor State, four KAP Surveys had been conducted among rice and cocoa farmers. In Penang State, three KAP Surveys were carried out among durian (Durio zibethinus) growers. Participants from the Kedah State also reported to have conducted KAP Surveys.

→ Pretesting of extension, training and communication materials has now become a standard practice for DOA's Extension Branch, especially for multi-media materials development activities of its Development Support Communication section.

→ A multi-media support approach to extension and training activities has been widely accepted among extension personnel of DOA who now avoid developing messages with general information or campaign slogans.

→ Special training of the "intermediaries" such as contact farmers, field extension workers, as well as other community-based volunteers such as local/religious leaders, school teachers, agricultural inputs/supplies retailers, etc., has been made an integral part of the extension strategies.

→ If the Core-Group members can be considered the "First Generation" SEC resource persons, as of 1989, as reported by Mohamed (1989), there were already three generations of such persons. The Second Generation of SEC specialists were "brought up" during the replication of a SEC process on Durian Fruit Borer (Plagideicta magniplaga) in 1988. Third Generation SEC specialists emerged when the "alumni" of the SEC on Durian Fruit Borer activities subsequently conducted a SEC replication on Synchronized Rice Planting Methods for farmers in Seberang Pantai in 1989. Another SEC on Pomelo (Limau Bali) Cultivation was implemented by the SEC Core-Group members, as shown in Fig. 4-19.

→ A major SEC replication in Malaysia was requested by Muda Agricultural Development Authority (MADA) to assist in its extension efforts to promote Integrated Weed Management in irrigated rice fields in north Malaysia. MADA requested FAO and DOA to assist in this SEC programme. Six members of the SEC Core-Group served as the principal trainers and planners of this SEC programme, and thus produced another Second Generation SEC specialists who are staff members of MADA. The SEC on Integrated Weed Management was as comprehensive and complete as that of the Rat Control, which included SEC training for 24 persons. The total cost of the activities was estimated at about US$ 46,400. More information on this activity will be provided later in Section 4.4. MADA extension and training staff have since carried out numerous SEC activities, and a recent (1992) campaign on Dry Seeding Method in Rice Cultivation.

→ There are indications that SEC has been formalized and institutionalized at some agriculture training centres and institutions of higher learning in Malaysia. Two training institutes and two universities whose staff participated in the SEC on Rat Control (thus members of the First Generation SEC Core-Group) have incorporated SEC into their curriculum structures :

● The DOA's Centre for Extension Development and Training in Telok Chengai, in Kedah State, since 1986 has included SEC as part of a two-week course on extension and training methodology.
● The Agricultural Training Institute in Bumbung Lima, in Penang State, started in 1989 to offer SEC as a course for extension officers.

FIGURE 4-19

Sustainability and Multiplier Effects of SEC Activities in Malaysia: Selected Examples

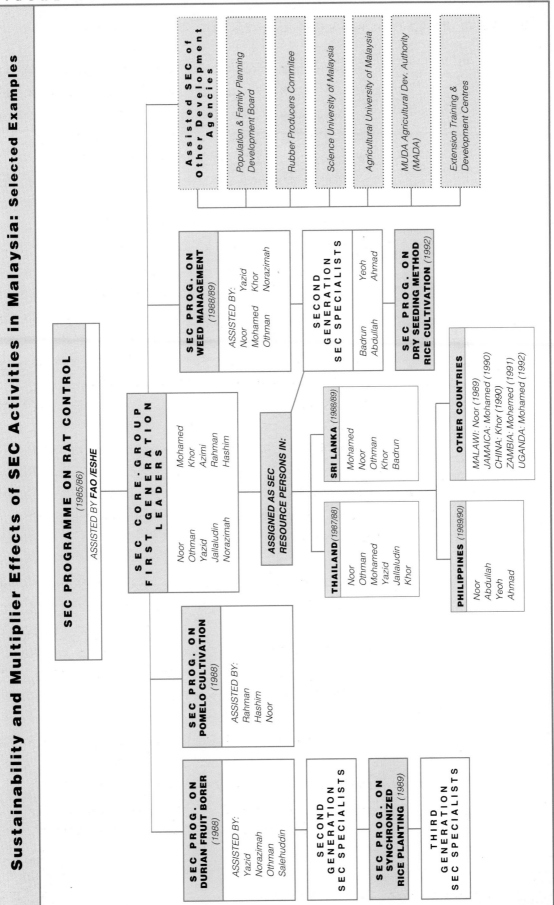

SEC PROGRAMME ON RAT CONTROL *(1985/86)*

ASSISTED BY FAO/ESHE

SEC CORE-GROUP FIRST GENERATION LEADERS
Noor
Othman
Yazid
Jallaludin
Norazimah
Mohamed
Khor
Azimi
Rahman
Hashim

SEC PROG. ON WEED MANAGEMENT *(1988/89)*
ASSISTED BY:
Noor   Yazid
Mohamed   Khor
Othman   Norazimah

SECOND GENERATION SEC SPECIALISTS
Badrun   Yeoh
Abdullah Ahmad

SEC PROG. ON DRY SEEDING METHOD RICE CULTIVATION *(1992)*

Assisted SEC of Other Development Agencies
Population & Family Planning Development Board
Rubber Producers Commitee
Science University of Malaysia
Agricultural University of Malaysia
MUDA Agricultural Dev. Authority (MADA)
Extension Training & Development Centres

SEC PROG. ON POMELO CULTIVATION *(1988)*
ASSISTED BY:
Rahman
Hashim
Noor

SEC PROG. ON DURIAN FRUIT BORER *(1988)*
ASSISTED BY:
Yazid
Norazimah
Othman
Salehuddin

SECOND GENERATION SEC SPECIALISTS

SEC PROG. ON SYNCHRONIZED RICE PLANTING *(1989)*

THIRD GENERATION SEC SPECIALISTS

ASSIGNED AS SEC RESOURCE PERSONS IN:

THAILAND *(1987/88)*
Noor
Othman
Mohamed
Yazid
Jallaludin
Khor

SRI LANKA *(1988/89)*
Mohamed
Noor
Othman
Khor
Badrun

PHILIPPINES *(1989/90)*
Noor
Abdullah
Yeoh
Ahmad

OTHER COUNTRIES
MALAWI: Noor (1989)
JAMAICA: Mohamed (1990)
CHINA: Khor (1990)
ZAMBIA: Mohemed (1991)
UGANDA: Mohamed (1992)

Note: this figure shows examples of some SEC replications in Malaysia assisted by Malaysian SEC-trained resource persons who have also served in other countries (as of Dec. 1992)

● The Science University of Malaysia offers two courses dealing with important SEC aspects, such as campaign planning and implementation (course YBP 302), and formative evaluation or pretesting of multi-media materials (course YBP 303).

● The Centre for Extension and Continuing Education (CECE), Agricultural University of Malaysia, also offers SEC-related two courses, covering campaign strategy development (course PP 431) and multi-media extension methods (course PP 307).

→   While the Core-Group members are actively sharing and disseminating their SEC skills/experiences at the national level, they have also served effectively as resource persons to train and help plan SEC replications internationally. As shown in Fig. 4-19, and also below, the following countries have benefited from Malaysian SEC specialists as of July 1993:

COUNTRY	No. of Malaysian SEC resource persons who had served there
THAILAND	6
SRI LANKA	5
PHILIPPINES	4
CHINA	1
MALAWI	1
ZAMBIA	1
UGANDA	1
JAMAICA	1

The SEC results in Malaysia to date provide ample evidence that SEC is being institutionalized. Therefore, SEC benefits to Malaysia are not limited to the economic gain resulted from the Rat Control Campaign, but also a useful contribution for further strengthening and improving the agricultural extension service and its system.

4.3.
Thailand:
the Pest Surveillance
System Campaign

In Thailand, a Strategic Extension Campaign to promote the appropriate application of a pest surveillance system among rice farmers was carried out in 1988. The Thai's Department of Agricultural Extension (DOAE) and the GTZ-funded Thai-German Plant Protection Programme (TGPPP) requested FAO to provide the necessary technical assistance in planning the SEC programme and training DOAE personnel in SEC process & methodology. The FAO's IPM project, in collaboration with the FAO's Agricultural Education and Extension Service (ESHE), provided the necessary technical inputs, mainly for the SEC planning, training, as well as KAP & other evaluation studies. In the spirit of the Technical Cooperation among Developing Countries (TCDC) concept, FAO utilized qualified SEC trainers or resource persons from the region (i.e., Malaysia and the Philippines) who were previously trained by FAO staff in SEC workshops in Malaysia. All these persons had first-hand experience in planning, implementing and managing one or more SEC programmes elsewhere, following the complete SEC process and its methodology.

The SEC on Pest Surveillance System, which involved the training of 25 Thai Core-Group members in five SEC workshops, was aimed at 12,000 farmers, 5,000 school children, and 400 local leaders/teachers in 116 villages in Chainat Province, Thailand. The detailed schedule of activities for this SEC programme is shown in Figure 4-20. This SEC programme followed closely the process and procedures applied to the Malaysia's SEC programme on Rat Control, including the operational steps and evaluation procedures. The cost of this SEC programme was estimated at about US$ 94,617.

The SEC activities included a series of 5 workshops, held between March 1987 - March 1988, to train 25 subject-matter specialists, extension officers and trainers of Thailand's Dept. of Agricultural Extension (DOAE) on various SEC concepts, principles and techniques, as well as their direct involvement in

Activities and Schedule for Strategic Extension Campaign on Pest Surveillance System in Chainat Province, Thailand

STEP	ACTIVITIES	DATE	RESOURCES NEEDED	PROVIDED BY
1	Workshop on rationale, purpose and study design for conducting a survey of Thai farmers' Knowledge, Attitude and Practice (KAP) regarding Rice Pest Surveillance System	11-17 March 1987	Consultant (international) for 10 m/days DSA: 7 days x 25 persons x B. 510 = B. 90,000 (= US $ 3,500)	FAO FAO
2	KAP survey of the Rice Pest Surveillance System among Thai farmers in the Central Districts	April-July 1987	Expenses for survey investigators from Chulalongkorn University (US $ 4,500)	FAO
3	Workshop on Extension Campaign Planning, Message Design and Materials Development (to prepare for the strategic multi-media campaign on the adoption of the Rice Pest Surveillance System by Thai Farmers in the Central Districts)	19-29 Oct. 1987	Resource Persons (international and national) for 15 days DSA: 15 days x 25 persons x B. 510 = B. 19,000 (= US $ 7,500) Travel expenses for 25 participants	FAO TGPPP TNCIPC/ DOAE
4	Workshop on Pretesting/Formative Evaluation of Prototype Campaign Materials (which have been developed during the October 1987 workshop mentioned above)	12-18 Nov. 1987	Consultant (international) for 7 m/days DSA: 6 days x 25 persons x B. 510 = B. 76,500 (= US $ 2,950) Travel expenses for 25 participants	FAO TGPPP TNCIPC/ DOAE
5	Pretesting prototype campaign materials	16-18 Nov. 1987	Part of step 4	
6	Reproduction of the extension campaign materials (whose prototypes have been pretested and revised during the November 1987 workshop mentioned above)	Nov. 1987 to Jan. 1988	Expenses for media materials production (US $ 25,000)	TGPPP
7	Workshop on Campaign Management Planning (to prepare for implementation of the campaign)	25-30 Jan. 1988	Consultant (international) for 8 m/days DSA: 6 days x 25 persons x B. 510 = B. 76,500 (= US $ 2,950) Travel expenses for 25 participants	FAO TGPPP TNCIPC/ DOAE
8	Training and orientation for campaign personnel	February 1988	Active involvement of DOAE personnel	TNCIPC/ DOAE

FIGURE 4-20 CONTINUED

Activities and Schedule for Strategic Extension Campaign on Pest Surveillance System in Chainat Province, Thailand (continued)

STEP	ACTIVITIES	DATE	RESOURCES NEEDED	PROVIDED BY
9	Workshop on Campaign Evaluation Methods and Management Monitoring Procedures	8-11 March 1988	Consultant (international) for 7 m/days	FAO
			DSA: 6 days x 25 persons x B. 510 = B. 76,500 (= US $ 2,950)	TGPPP
			Travel expenses for 25 participants	DOAE
10	Implementation of Strategic Multi-Media Campaign on the Adoption of Rice Pest Surveillance System by Thai Farmers in the Central Districts	April-Nov. 1988	Active involvement of DOAE personnel	TNCIPC/ DOAE
11	Evaluation Studies: ● Management Monitoring Survey (MMS)	May-June 1988	US $ 3,500	FAO
	● Information Recall and Impact Survey (IRIS) and Focus Group Interviews (FGI)	Sept.1988 & May '89	US $ 8,500 for both studies	TGPPP/FAO
	● Field Damage Assessment Study	Nov. 1988- Jun. 1989	By DOAE plant protection personnel (no additional cost involved)	TNCIPC/ DOAE
12	International Seminar on : Experience Sharing and Results Dissemination regarding the Campaign Planning, Implementation and Evaluation	August 1989	DSA and travel for 40 local participants; 8 invited international participants (DSA & travel)	DOAE TGPPP/FAO
			2 international resource persons for 10 m/days each	FAO
			Publication/Documentation of Campaign process and results (US $ 8,000)	FAO

Note:
FAO here refers to FAO projects GPC/RAS/108/AGF, GCP/RAS/101/NET, GCP/RAS/092/AUL, assisted by the FAO's Agricultural Education and Extension Service (ESHE) staff.
TGPPP = Thai-German Plant Protection Programme
TNCIPC/DOAE = Thai National Committee for Integrated Pest Control/Department of Agricultural Extension
DSA = Daily Subsistence Allowance
B = Thai Baht (US$ 1 = B 25.50 based on excange rate in 1989)

actual field action and follow-up activities as part of the campaign planning, implementation, management and monitoring process. The Campaign was launched in April 1988 (before the rice planting season) and terminated in November 1988 (just before harvest).

Evaluation studies had specifically been commissioned to individuals or institutions (i.e., Chulalongkorn and Kasetsart universities and Suwannaphum Agro Consultants) not involved in the campaign planning or implementation process in order to obtain objective assessment of the effectiveness and usefulness of the SEC on Pest Surveillance System in Thailand. The following types of evaluation studies were conducted :

BEFORE Campaign[1]	DURING Campaign[2]	AFTER Campaign[3]
Knowledge, Attitude & Practice (KAP) Survey, incl. Focus Group Interviews (FGI)	Management Monitoring Survey (MMS)	Information Recall & Impact Survey (IRIS), including FGI

In the following pages, a summary of the campaign's evaluation results and also information on the important elements/aspects of this SEC process and method are provided.

Details on the results of the evaluation studies are contained in the following reports:

1 T. Boonlue, "Knowledge, Attitude, and Practice on Pest Surveillance System in Chainat Province, Thailand", Bangkok: Chulalongkorn Univ., October 1987
2 P. Boonruang and P. Chunsakorn, Management Monitoring Survey of the Pest Surveillance Campaign on Rice", Bangkaen: Kasetsart Univ., June 1988
3 C. Tiantong, "An Information Recall and Impact Survey (IRIS) on the Strategic Extension Campaign on Pest Surveillance System in Rice, in Chainat Province, Thailand", October 1988

STRATEGIC EXTENSION CAMPAIGN (SEC)
ON PEST SURVEILLANCE SYSTEM
IN THAILAND

GENERAL OBJECTIVE: TO PROMOTE APPROPRIATE APPLICATION OF PEST SURVEILLANCE SYSTEM AMONG RICE FARMERS (for specific and measurable objectives see page 137)

DURATION: APRIL - NOVEMBER 1988

TARGET LOCATION: 17 SUB-DISTRICTS / 116 VILLAGES IN CHAINAT PROVINCE, THAILAND

TARGET AUDIENCE: about 12,000 FARMERS
5,000 SCHOOL CHILDREN
400 LOCAL LEADERS & TEACHERS

ESTIMATED COST: US $ 94,617 (for one rice planting season), including the one-time human resources development investment for the initial training of 25 extension staff on SEC methods

EVALUATION PROCEDURES:
- KNOWLEDGE, ATTITUDE AND PRACTICE (KAP) SURVEY
- MANAGEMENT MONITORING SURVEY (MMS)
- INFORMATION RECALL AND IMPACT SURVEY (IRIS)
- FOCUS GROUP INTERVIEWS (FGI) (for qualitative aspects of KAP and IRIS)

Campaign Strategy Planning Workshop

FIGURE 4-21

Identified Problems of
Pest Surveillance System
Based on Farmers' KAP Survey in Chainat Province, Thailand

	IDENTIFIED PROBLEM	PROBLEM RELATED TO
1	Low knowledge on pest identification and Economic Threshold Level (ETL)	KNOWLEDGE
2	Lack of sufficient knowledge on the importance and benefits of natural enemies	KNOWLEDGE
3	Lack of sufficient knowledge on the importance and benefits of resistant rice varieties	KNOWLEDGE
4	Lack of awareness on Surveillance and Early Warning System (SEWS) programme, and ability in using Pest Surveillance form	KNOWLEDGE/ PRACTICE
5	Farmers prefer broad-spectrum pesticides and blanket spraying	ATTITUDE
6	Farmers do not believe in the effectiveness of natural enemies	ATTITUDE
7	Farmers go to the edge of the field, but NOT into the field to check for pests according to the recommended precedure and frequency	ATTITUDE
8	Farmers spray pesticides on sight of pests based on their "natural instinct"	ATTITUDE
9	Farmers are aware of pesticide hazards, but DO NOT apply safety precautions in pesticide handling, application and disposal	PRACTICE

FIGURE 4-22

Specific and Measurable Campaign Objectives
Based on the Problems Identified by the KAP Survey for the Strategic Extension Campaign (SEC) on Pest Surveillance System in Chainat Province, Thailand

	IDENTIFIED PROBLEMS	EXTENSION CAMPAIGN OBJECTIVES

1	Low knowledge on pest identification and necessary action for pest control	To increase the percentage of farmers who have knowledge regarding:
		  a. Pest identification from 41% to 65% and,
		  b. Necessary action for pest control from 15.1% to 40%

2	Lack of sufficient knowledge on the importance and benefits of natural enemies	To increase the percentage of farmers who know the identity of natural enemies (good bugs) from 11.4% to 35%
3	Lack of sufficient knowledge on the importance and benefits of resistant rice varieties	To increase the percentage of farmers who have knowledge regarding the recognition and importance of resistant rice varieties from 35.8% to 50%
4	Lack of awareness on Surveillance and Early Warning System (SEWS) programme and Pest Surveillance (PS) form	To create awareness by increasing the percentage of farmers having knowledge on SEWS from 13.2% to 50% and to increase the percentage of farmers skilled in the use of Pest Surveillance (PS) form from 10.1% to 30%

5	Farmers prefer broad-spectrum pesticides and blanket spraying	To reduce the percentage of farmers using broad-spectrum pesticides by:
		  a. Increasing the percentage of farmers who know how to choose right chemicals from 5% to 16%
		  b. Decreasing the percentage of farmers who prefer broad-spectrum pesticides from 65% to 50%

6	Farmers do not believe in the effectiveness of natural enemies	To reduce the percentage of farmers who do not believe that conservation of natural enemies can suppress pest population from 36.5% to 25%
7	Farmers go to the edge of the field, but NOT into the field to check for pests according to the recommended precedure and frequency	To increase the percentage of farmers who check their fields according to the recommended procedure from 17% to 35%
8	Farmers spray pesticides on sight of pests based on their "natural instinct"	To reduce the percentage of farmers who believe in the need for spraying pesticides as soon as pests are observed in the field, without checking the field properly, from 69.8% to 55%
9	Farmers are aware of pesticide hazards, but DO NOT apply safety precautions in pesticide handling, application and disposal	To increase the number of farmers observing adequate safety measures in using pesticides by increasing the percentage of farmers practising correct disposal of left-over pesticide from 10.7% to 25%

FIGURE 4-23

Strategy Development and Message Design Worksheet

	IDENTIFIED PROBLEMS	REASONS FOR PROBLEMS	PROBLEM SOLVING STRATEGY	MESSAGE APPEAL	EXAMPLE OF MESSAGE	CHANNEL FOR MESSAGE DELIVERY
1	Low knowledge on pest identification and Economic Threshold Level (ETL)	1. Lack of Plant Protection Service Unit (PPSU) officers to train farmers directly	1.1 Extension Agents (EA) will assist PPSU officers to train farmers	Morale boosting	"Contact your nearest PPSU officer or EA who are skillful and trained in pest surveillance"	All materials
		2. EA have low knowledge on pest surveillance and Surveillance and Early Warning System (SEWS)	2.1 PPSU officers will train EA on pest identification and surveillance system			Flipchart, booklet, Pest Surveillance (PS) form
		3. Lack of effective training materials	3.1 Develop training and reference materials suitable for EA and farmers	Technology Simplification		Flipchart, leaflet
		4. Poorly trained farmers	4.1 Same as below	Testimonial/ Informational	Farmer who owns field explaining how crop losses can be avoided by proper checking and control measures	Leaflet, farmer-to-farmer, video, poster, booklet, PS form
			4.2 Encourage farmers to visit field trials			
			4.3 Increase farmers' knowledge on pest identification - emphasis on planthopper			
2	Lack of sufficient knowledge on the importance and benefits of natural enemies (NE)	1. Lack of PPSU officers to train farmers	1.1 EA will assist PPSU officers to train farmers			
		2. EA not very familiar with natural enemies	2.1 PPSU officers will train EA on NE (emphasis on spider)			
		3. Farmers have not seen how NE attack pests in the field	3.1 Show the real action in the field	Incentive	"Spiders (NE) are useful, and can be your assistant"	Booklet; poster Video, flyer, flipchart, radio spot, song, leaflet audio-cassette
				Dramatized approach	"Spiders can kill and eat planthoppers better than we can"	

FIGURE 4-23 (CONTINUED)

Strategy Development and Message Design Worksheet (continued)

No.	Problem	Objective	Approach	Message	Media
3	Lack of knowledge on importance and benefits of resistant rice varieties	1. Lack of knowledge on resistant rice varieties to specific pests			
		1.1 Demonstrate plots of resistant versus non-resistant varieties	Incentive/Reward approach	"No pest problems when using resistant varieties"	Booklet, video, leaflet
		1.2 Provide more information on resistant varieties	Testimonial approach	Interviews with farmers who have planted resistant varieties	Flyer, flipchart, radio spot, song, audio-cassette
4	Lack on Surveillance and Early Warning System (SEWS) programme and ability to use Pest Surveillance form	1. Limited exposure to program and use of Pest Surveillance (PS) form			
		1.1 Increase exposure through use of mass media	Social prestige approach	"Modern farmers participate to SEWS program"	Billboard, radio spot, song, audio-cassette, sticker, poster, flipchart, video
			Economic/Reward Incentives	"Farmers in SEWS area get more benefits"	
		2. Pest Surveillance (PS) form is too complicated for farmers: - too much information is requested - not practical for field use - easily damaged by moisture in the field			
		2.1 Introduce a new, simple, easy to carry, and attractive PS form	Simplification approach	"The form helps: - identify correct pests - make right decision before spraying"	Pocket-size PS form
		2.2 Stress importance and advantages of the PS form for deciding farmers' pest control action	Incentive/Reward approach		Video, radio spot, song, audio-cassette, flipchart, leaflet, sticker, billboard
5	Farmers prefer broad-spectrum pesticides and blanket spraying	1. Broad-spectrum pesticides are preferred because: - cheap (save money) - multipurpose (no need to buy other chemicals) - easily available			
		1.1 Emphasize bad effects of broad-spectrum pesticides: - kill both pests and NE - contaminate environment	Fear arousal	"Broad-spectrum chemicals are poisonous"	Booklet, video, leaflet
			Down playing the competitor	"Better not to spray (thus save money) than to use broad-spectrum chemicals"	Flipchart, video
		1.2 Demonstrate advantages of spot-spraying	Incentive/Reward approach	"Spot-spraying is cheaper and more effective"	Flipchart, video
				"Spot-spraying is safer for the environment, NE and human beings"	Booklet, PS form, flyer

FIGURE 4-23 (CONTINUED)

Strategy Development and Message Design Worksheet (continued)

	IDENTIFIED PROBLEMS	REASON FOR PROBLEMS	PROBLEM SOLVING STRATEGY	MESSAGE APPEAL	EXAMPLE OF MESSAGE	CHANNEL FOR MESSAGE DELIVERY
6	Farmers do not believe in the effectiveness of natural enemies	1. Farmers have not seen how natural enemies destroy pests	1.1 Show the real action in the field			Field demonstration, TV, video
			1.2 Mass informatiion/education on usefulness of NE which are to be released in large number in the field at campaign opening	Patronage approach	Some influencial person releasing the reared NE during the campaign launching day	TV, radio, newspaper
			1.3 Encourage conservation of spiders	Incentive/Reward approach	"One spider can kill 14 planthoppers per day"	Booklet, flipchart, flyer, poster
					"Use less pesticides and save money"	Booklet, flipchart
				Testimonial approach	Successful farmers explaining benefits and effectiveness of NE	Video
7	Farmers do not go into the field to check for pests	1. Rice plants might be damaged	1.1 Demonstrate actual results/evidence that rice plants are nor damaged if field checking is done carefully	Factual demonstration approach	"Walking into the field to check for pests does not affect yield"	Field demonstration , video, booklet, flip-chart
				Ridicule approach	"Improper checking (from the dike) wastes time & effort	Poster, comic sheet
		2. It is inconvenient and time consuming	2.1 Introduce simplified methods and procedures	Simplicity approach	"Going into the field to check for pests is easy and effective"	Sticker, flipchart, video, leaflet,PS form

FIGURE 4-23 (CONTINUED)

Strategy Development and Message Design Worksheet (continued)

			Emotional approach		
8	Farmers spray pesticides when they see pests in the field, based on their "natural instinct"	1. Farmers are used to this method	1.1 Increase awareness on chemical toxicity	"Decrease use of chemicals and increase your life span" Fear arousal — "Beware - do not use chemical excessively, your family's health is in danger"	Poster Poster, flipchart, video
		2. Avoiding risk of spreading the infestation to other areas	2.1 Increase knowledge on - concept of surveillance - natural enemies - ETL	Simplicity of procedures — "Check the fields and count for planthoppers on 4 hills. If more than 40 planthoppers on 4 hills and no spiders, then only spray"	Booklet, video, leaflet, form
9	Farmers are aware of pesticide hazards BUT DO NOT observe safety precautions in pesticide handling, application and disposal	1. Farmers believe that some people are immune to chemical toxicity, so they can spray without protection	1.1 Show cases of pesticide poisoning and consequent effects on family life	Fear arousal — "Improper use of pesticides can kill you and members of your family"	Video, leaflet
		2. Wearing protective clothing is too hot	2.1 Same as below	Fear arousal — "Many people died of pesticide poisoning. Who will be next?"	
		3. There are convenient places for disposal of empty pesticide containers and left-over pesticides	3.1 Show effects of improper disposal of containers and left-over pesticides	Guilt feeling — "Disposal of containers and left-over pesticide into canals can cause poisoning to your friends and neighbours "	Booklet, flipchart
			3.2 Encourage proper handling of pesticides	Group pressure — "Don't dispose empty containers and left-over pesticides into canals, because it may harm your neighbours and friends"	Poster
		4. There is a market for empty containers	4.1 Discourage chemical manufacturers to recycle empty containers		

FIGURE 4-24

Strategic Plan for Multi-Media Campaign on Pest Surveillance System in Chainat Province, Thailand

FARMERS

No SEWS Training; Age 26-35 (21%); 36-55 (52%);
Male (63%), Female (37%); Education 1-4 years (87%)

Department of Cooperatives

- Meeting/Lobbying
- Flyer (P-14)
- Endorsement/Legitimization

Education Department

- Comic Sheet (P-13)
- Meeting/Lobbying
- Endorsement/Legitimization

District Agric. Coop Officers

- Meetings
- Flyer

Members of Agric. Coop

- Interpersonal/Group Discussion

School Teachers

School Children

- Family Discussion

- Village Newspapers
- Reading Centers
- Sub-section
- Irrigation Stations

- Pesticide Dealer Shops
- Community Centers
- Village Coffee Shops
- District Offices

- Farmers' Centers
- Cooperatives
- District BAAC
- Markets

Subject Matter Specialists (SMS)

- Motivational Posters (P-1, P-2, P-4)
- Booklet (P-3)
- Flipchart (P-5)
- Video (P-6)

Extension Agent (Kaset Tambon)

- Cloth Flipcharts (P-15)
- Training, Group/Interpersonal Discussion
- Pest Surveillance Form
- Training

Plant Protection Service Unit (PPSU) Staff

- Pest Surveillance Form (P-8)
- Leaflet (P-7)
- Audio-cassette (P-9)
- Training

Successful Farmers

- Interpersonal Discussion

Broadcasting Towers

- Sticker (P-11)

- Villages Gas Stations
- Mobile Units
- Official Vehicles

- Travelling Salesmen
- Farmers' Motorcycles
- K. Tambon Motorcycles

- Radio spots/Song (P-10)

- Billboards (P-12)

- Village Entrances

CAMPAIGN CENTER

Distribution/display means or points

Intermediaries

Delivery Agents/Channels

Media · Message · Audience Checklist

NO	MEDIA TYPE	MAIN MESSAGE	TO SOLVE PROBLEM	FOR WHOM
P-1	Motivational poster A	1. Spiders kill planthoppers 2. Excessive use of pesticides will destroy spiders in your field	1, 2, 6, 7	Farmers
P-2	Motivational poster B	1. Check your field planthoppers and NE (record on form): spray only if you find 40 hoppers in 4 plants or hills, and no spiders	4, 7, 8	Farmers
P-3	Booklet	1. Identifying pest and NE 2. Simplified technology on pest surveillance 3. Use of resistant varieties 4. Safe use of pesticides	1 - 9	PPSU staff Extension Agents
P-4	Motivational poster C	1. Safe use of pesticides	1 - 9	Farmers
P-5	Flipchart	1. Use of resistant varieties, identification of pest and natural enemies, steps in pest surveillance, safe-use of pesticides	1 - 9	SMSs PPSU staff Extension Agents
P-6	Video	1. Identification of natural enemies 2. Surveillance and Early Warning System 3. Safe use of pesticides	1 - 9	PPSU staff Extension Agents
P-7	Leaflet	1. Steps in pest surveillance 2. Use of resistant varieties	1 - 9	Farmers
P-8	Pest Surveillance (PS) form	1. Use of simplified PS form 2. Importance of checking fields properly 3. Use correct Economic Threshold Level 4. Use right chemicals	1, 5, 7, 8	Extension Agents Farmers
P-9	Audio-cassette tape	1. Motivational radio spots and songs	1 - 9	Extension Agents Farmers
P-10	Radio spots and songs	1. What is pest surveillance (SEWS)? For more information, ask for training 2. Proper disposal of pesticide containers 3. Spray only at correct ETL 4. What natural enemies can do 5. Use of resistant varieties 6. Going into the field is easy, will not destroy plants, and will not reduce yields	1 - 9	Farmers SMSs PPSU staff Extension Agents
P-11	Sticker	1. Motivation to go into the fields to check	4, 7	Farmers
P-12	Billboard	1. Motivation to go into the fields to check	4, 7	Farmers
P-13	Comic sheet	1. Result of farmer checking field from dike only (motivational story)	7	School children
P-14	Flyer	1. Identifying natural enemies 2. Use of resistant varieties	2, 3, 5, 6	Farmers
P-15	Cloth flipchart	1. Steps in a simplified technology in pest surveillance	1 - 9	Farmers

FIGURE 4-26

Campaign Inputs

MULTI-MEDIA MATERIALS / INPUTS		QUANTITY
P - 1; 2; 4	Poster (3 topics)	6,000
P - 3	Booklet	1,000
P - 5	Flipchart	250
P - 6	Video	25
P - 7	Leaflet (3 topics)	37,500
P - 8	Pest Surveillance (PS) form	15,000
P - 9; 10	Audio-cassette tape containing radio spots & songs	750
P - 11	Sticker	5,000
P - 12	Billboard	10
P - 13	Comic sheet	5,000
P - 14	Flyer	10,000
P - 15	Cloth flipchart	700

PERSONAL COMMUNICATION/ INFLUENCE INPUTS	SPECIALLY TRAINED	USED AS TRAINERS/ RESOURCE PERSONS	USED AS INTERME- DIARIES	PROVIDED WITH MULTI- MEDIA MATERIALS
Subject-matter specialists/ Plant protection officers	X	X		X
Extension workers	X	X		X
Local leaders	X	X		X
Teachers	X	X		X
School children			X	X
Cooperative members			X	X

SPECIAL EVENTS	REMARKS
Campaign Inauguration Ceremony	Conducted at Province, District, Sub-district, and Village Levels
Straw Model Contest and Floats (on Pest and their Natural Enemies)	Participated by about 30 Village Groups
Folk Dances and Singing Contest	Participated by representatives of about 20 Village Groups

FIGURE 4-27

Campaign Expenditures (in US $)

STAFF TRAINING $ 24,850

25 extension staff trained in 5 different workshops
totalling 36 days
(for DSA and travel expenses)

MATERIALS DEVELOPMENT & PRODUCTION $ 38,267

	TYPE	QUANTITY	UNIT COST $	TOTAL COST$
P - 1; 2; 4	Poster (3 topics)	6,000	(0.57)	3,440
P - 3	Booklet	1,000	(2.80)	2,805
P - 5	Flipchart	250	(27.76)	6,940
P - 6	Video	25	(80.00)	2,000
P - 7	Leaflet (3 topics)	37,500	(0.05)	1,920
P - 8	Pest Surveillance (PS) form	15,000	(0.37)	5,600
P - 9; 10	Audio-cassette tape containing radio spots & songs	750	(1.07)	800
P - 11	Sticker	5,000	(0.24)	1,200
P - 12	Billboard	10	(240.00)	2,400
P - 13	Comic sheet	5,000	(0.07)	360
P - 14	Flyer	10,000	(0.19)	1,936
P - 15	Cloth flipchart	700	(12.67)	8,866

SURVEYS & EVALUATION STUDIES $ 15,000

KAP survey (Baseline)
Management Monitoring Survey (MMS)
Informational Recall and Impact Survey (IRIS)

INTERNATIONAL RESOURCE PERSONS $ 16,500

For travel, DSA and fees

TOTAL $ 94,617

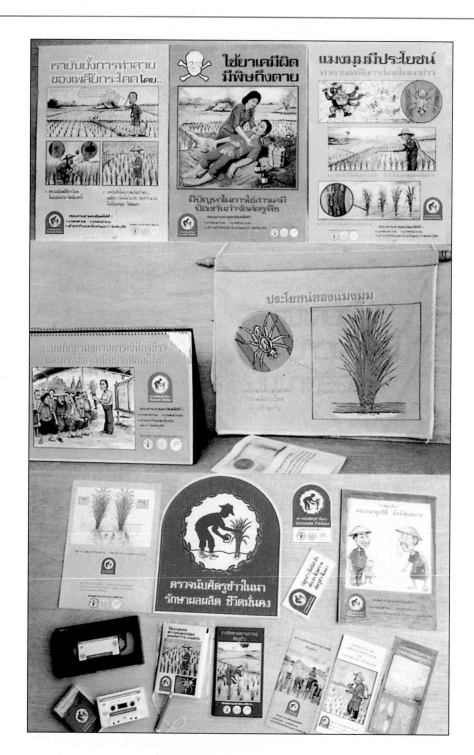

*Assorted Multi-Media Materials Used for the Strategic Extension Campaign
on Pest Surveillance System in Thailand*

Straw Models and Floats

of Pests and their Natural Enemies

Exhibited during

Campaign Inauguration

FIGURE 4-28

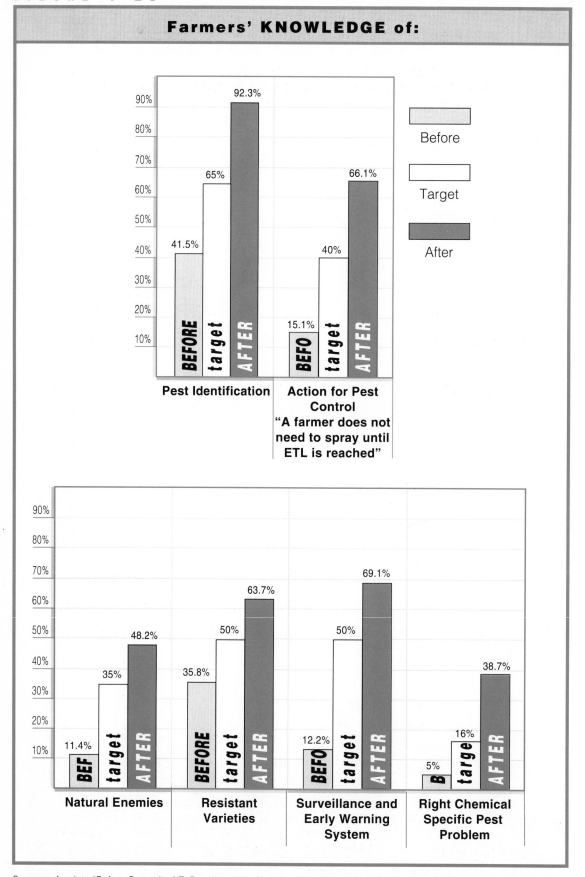

Farmers' KNOWLEDGE of:

Pest Identification
BEFORE 41.5% / target 65% / AFTER 92.3%

Action for Pest Control
"A farmer does not need to spray until ETL is reached"
BEFORE 15.1% / target 40% / AFTER 66.1%

Natural Enemies
BEFORE 11.4% / target 35% / AFTER 48.2%

Resistant Varieties
BEFORE 35.8% / target 50% / AFTER 63.7%

Surveillance and Early Warning System
BEFORE 12.2% / target 50% / AFTER 69.1%

Right Chemical Specific Pest Problem
BEFORE 5% / target 16% / AFTER 38.7%

Legend: Before / Target / After

Sources: for data "Before Campaign" T. Boonlue, 1987; for data "After Campaign": C. Tiantong, 1988

Farmers' ATTITUDE towards

Broad-Spectrum Pesticides:
"Use of broad-spectrum pesticides is the most effective method to control pests"

	Agree
	Not sure
	Disagree

Spraying Pesticides:
"Spray immediately when pests are observed in the field"

	Agree
	Not sure
	Disagree

Sources: for data "Before Campaign" T. Boonlue, 1987; for data "After Campaign": C. Tiantong, 1988

FIGURE 4-30

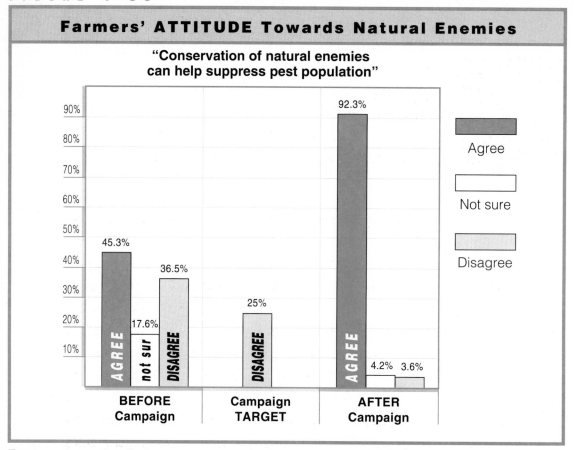

Farmers' ATTITUDE Towards Natural Enemies

"Conservation of natural enemies can help suppress pest population"

FIGURE 4-31

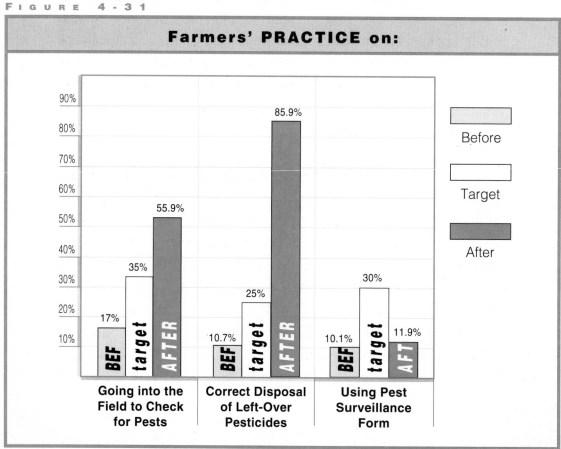

Farmers' PRACTICE on:

Sources: for data "Before Campaign" T. Boonlue, 1987; for data "After Campaign": C. Tiantong, 1988

4.4.
Malaysia:
the Integrated Weed
Management Campaign

The Strategic Extension Campaign (SEC) on integrated weed management in the Muda irrigation scheme was carried out by the Muda Agricultural Development Authority (MADA) which is a large irrigated agricultural development area scheme in North Malaysia. It is not part of Malaysia's Department of Agriculture (DOA), and it has its own Extension & Training Department. Upon learning of the successful results of the SEC on Rat Control conducted by the DOA, MADA requested FAO to assist in planning and conducting an SEC on Integrated Weed Management (IWM). MADA senior management officials after discussions with FAO staff agreed on a collaborative effort to implement this SEC programme to handle the weed problem which was fast becoming a serious one in the Muda irrigation scheme with the wide-spread implementation of direct-seeding method or rice planting. FAO's Agricultural Education and Extension Service (ESHE) and FAO's Inter-Country Programme on IPM in Rice (projects GCP/RAS/101/NET and GCP/RAS/092/AUL) provided the technical assistance in planning and designing the necessary SEC operational steps, and in providing SEC training support. The evaluation studies of the SEC on IWM were conducted by researchers from the Science University of Malaysia (USM).

One of the objectives of the SEC on IWM was to train a core-group of MADA staff on the process and skills of designing a well-planned and systematic extension and training programme based on the SEC principles. In addition to FAO staff, six of Malaysia's Dept. of Agriculture (DOA) staff members who were trained during the FAO-assisted SEC on Rat Control in Malaysia were the main resource persons in planning the SEC activities on IWM, and in providing SEC training to MADA personnel. As shown in Figure 4-32, SEC workshops and follow-up activities were conducted consistent with implementation steps suggested in Fig. 2-2. The first preparatory and training activity of the SEC on IWM started in October 1987, but the actual implementation of the campaign itself started only in

January 1989 and lasted through the dry planting season of 1989 (January - September). The campaign was undertaken in Districts III and IV of the Muda scheme with a target beneficiaries of 30,000 farm families. The effectiveness of the SEC activities was evaluated in November 1989.

As reported by Mohamed and Khor (1990) in their evaluation of the SEC on IWM, the campaign's total expenditure was US$ 46,400 (see Fig. 4-46). Evaluation studies also revealed that rice production yield in the campaign areas increased by about 9,500 tons, equivalent to US$ 2.33 million. The estimated cost-benefit ratio is 1:50, and the economic benefit per farm family who adopted the campaign recommendations is about US$ 195 per season (see fig. 4-45).

Mohd. Noor and Othman (1992) reported that in subsequent years, other strategic extension campaigns were launched by MADA without external assistance. Another IWM campaign was conducted for Districts I and II during the off-season period (February - July) in 1990. The following year (1991) another SEC campaign was undertaken focusing on the Importance of Following the Planting Schedule as determined by MADA. In 1992, an SEC was conducted that emphasized the correct planting techniques of Dry Seeding Method of Rice Cultivation. These two campaigns covered all four MADA districts.

Ho (1994) reported that as a result of the Campaign, the infestation of Echinochloa crusgalli and E. colona was reduced by 66 percent in 1989/90, whilst rice yield increased by 27 percent, as compared to the 1988 season prior to the campaign. Ho (1994) also pointed out that "...Continuous implementation of SEC on IWM in the period of 1990-1993 over the entire Muda area has shown remarkable results. The dry-seeded first season rice yields have increased steadily, reaching 4.2 tons/ha. in 1993. Meanwhile the wet seeded second season rice yields have consistently been above 5.0 tons/ha..It is noteworthy that over the same period, the usage of herbicides has declined." (see Fig. 4-47).

In the following pages, some important information on the planning, implementation process, and results of the SEC on Integrated Weed Management (IWM) in 1993. Meanwhile the wet seeded second season rice yelds have consistently been above 5.0 tons/ha. conducted by MADA in 1989 are provided.

STRATEGIC EXTENSION CAMPAIGN (SEC) ON INTEGRATED WEED MANAGEMENT IN THE MUDA IRRIGATION SCHEME, MALAYSIA

GENERAL OBJECTIVE: MOTIVATE AND EDUCATE FARMERS ON THE PROPER USE OF CHEMICAL AND NON-CHEMICAL METHODS OF WEED CONTROL (for specific campaign objectives, see Figure 4-34)

DURATION: JANUARY 1989 - SEPTEMBER 1989

TARGET LOCATION: DISTRICTS III AND IV OF THE MUDA IRRIGATION SCHEME, MALAYSIA

TARGET AUDIENCE: 30,000 FAMILIES

ESTIMATED COST: US $ 46,409

EVALUATION PROCEDURES: KNOWLEDGE, ATTITUDE AND PRACTICE (KAP) SURVEY FOCUS GROUP INTERVIEW (FGI) MANAGEMENT MONITORING SURVEY (MMS) INFORMATION RECALL AND IMPACT SURVEY (IRIS)

FIGURE 4-32

Activities and Schedule for SEC on Integrated Weed Management in Muda Irrigation Scheme, Malaysia

STEP	ACTIVITIES	PROPOSED DATES/TIMES	MAJOR RESOURCES NEEDED TO UNDERTAKE ACTIVITY type of resources/inputs	to be provided by
1	Workshop on the rationale, purpose and study design for conducting a survey of Muda's Farmers' Knowledge, Attitude, and Practice (KAP) regarding Integrated Weed Management	October 1987	Consultant (national) for 10m/days DSA and travel for participants	FAO * MADA
2	(KAP) Survey on Integrated Weed Management	October 1987- January 1988	Contract	FAO
3	Workshop on Extension Campaign Planning, Message Design and Materials Development (to prepare the strategic multi-media campaign on Integrated Weed Management)	1 - 13 February 1988	Resource Persons (national) for 15 days DSA and travel expenses for 28 participants **	FAO MADA
4	Workshop on Pretesting/Formative Evaluation of Prototype Campaign Materials developed during the Feb. 1988 Workshop mentioned above	April 1988	Consultants (national) for 7m/days DSA and travel expenses for 28 participants	FAO MADA
5	Pretesting Prototype Campaign Materials	2 days in April 1988	Part of Step 4	
6	Reproduction of the Extension Campaign Materials	3 months	Contract to local media materials production firms	MADA
7	Workshop on Campaign Management Planning (to prepare for the implementation of the above campaign)	9 - 13 August 1988	Consultant for 8m/days DSA and travel expenses for 28 participants Consultant (national) for 7m/days	FAO MADA FAO
8	Training and Orientation for Campaign Personnel	August - December 1988	Active involvement of MADA personnel	MADA

FIGURE 4-32 CONTINUED

Activities and Schedule for SEC on Integrated Weed Management in Muda Irrigation Scheme, Malaysia (continued)

STEP	ACTIVITIES	PROPOSED DATES/TIMES	MAJOR RESOURCES NEEDED TO UNDERTAKE ACTIVITY Type of resources/inputs	to be provided by
9	Workshop on Campaign Evaluation Methods and Management Monitoring Procedures	15 - 17 August 1988	Consultant (national) for 7m/days	FAO
			DSA and travel expenses for 28 participants (can be made part of Step 7)	MADA
10	Implementation of Strategic Multi-Media Campaign on the Integrated Weed Management	January - September 1989	Active involvement of MADA personnel	MADA
11	Evaluation Studies:			
	a. Management Monitoring Survey (MMS)	February 1989	Contract	FAO
	b. Information Recall and Impact Survey (IRIS)	November 1989	Contract	FAO
	c. Field Damage Assessment Survey (FDAS)	November 1989	By MADA plant protection personnel (no additional cost involved)	MADA
12	International Seminar on: Experience Sharing and Results Dissemination regarding the Campaign Planning, Implementation and Evaluation	June 1990	DSA and travel expenses for 40 local participants	MADA
			8 invited international participants (DSA&Travel)	FAO
			2 national resource persons for 10 m/days each	FAO
			Publication/documentation of Campaign process and results (US $ 4,000)	FAO/MADA

Notes:
* FAO here refers to FAO projects: GCP/RAS/101/NET and for GCP/RAS/092/AUL
** The participants of all the workshops were the same participants who attended the first Workshop in October 1987

FIGURE 4-33

The Strategic Extension Campaign on Integrated Weed Management in the Muda Irrigation Scheme, Malaysia

IDENTIFIED PROBLEMS

1. Farmers misconceived that rats are more dangerous than weeds

2. Farmers misconceived that weed control is a waste of time, money and effort

3. Farmers have little knowledge on how to identify barnyard grass and Leptochloa grass at two-leaf stage

4. Farmers did not use herbicides to control weeds due to its high costs

5. A large number of farmers did not use the right herbicide to control weeds

6. Farmers did not apply herbicide at the recommended rate and time

7. A low percentage of farmers practise proper cultural methods to control weeds especially in dry rotovation, land levelling, and selection of grass seeds

Note: the above mentioned problems were identified and prioritized based on the results of the survey of farmers in the MUDA Irrigation Scheme on their knowledge, attitude and practice (KAP) regarding Integrated Weed Management. The KAP Survey was conducted by Mohamed and Khor (1988).

FIGURE 4-34

The Strategic Extension Campaign on Integrated Weed Management in the Muda Irrigation Scheme, Malaysia

OBJECTIVES OF CAMPAIGN

Problems	Objectives
1. Misconception among farmers that rats are more dangerous than weeds	To increase the proportion of farmers who perceive weeds as a major pest from 24% to 45%
2. Misconception among farmers that controlling weeds is a waste of time, money and effort	To decrease the proportion of farmers who consider that: a. Controlling weeds is a waste of money and effort from 33% to 27% b. Hand weeding as a waste of time from 92% to 80%
3. Low knowledge on identification of barnyard and Leptochloa grass at 2-leaf stage	To increase the proportion of farmers who can identify the two most important weeds at 2-leaf stage: a. Barnyard grass from 55% to 65% b. Leptochloa grass from 75% to 78%
4. Low usage of herbicides by farmers to control barnyard and Leptochloa grass due to its high cost	To reduce the proportion of farmers who do not use herbicide to control barnyard and Leptochloa grass due to high cost from 57% to 37%
5. Farmers did not use the right herbicide to control barnyard and Leptochloa grass	To increase the proportion of farmers using recommended herbicides: a. STAM F-34 from 8% to 14% b. ARROSOLO from 5% to 9%
6. Farmers did not apply herbicide at the recommended rate and time	To increase the proportion of farmers knowledge on the proper use of herbicides according to recommended rate and time a. Rate/dosage: STAM F-34 from 9% to 19% ARROSOLO from 9% to 19% b. Time: STAM F-34 from 11% to 28% ARROSOLO from 30% to 60%
7. Farmers do not practise proper cultural methods to control weeds particularly in: a. dry rotovation b. land levelling c. selection of good seeds	To increase the proportion of farmers practising proper cultural methods to control weeds: a. Dry rotovation (twice) from 25% to 60% b. Land levelling after rotovation from 15% to 30% c. Selection of seeds from 59% to 75%

→ **AN EXAMPLE OF
SIMPLIFIED TECHNOLOGY RECOMMENDATIONS
FOR CONTROL OF BARNYARD GRASS (*Echinocloa
species*) AND RUMPUT MIANG (*Leptochloa chinensis*) IN DRY-SEEDED RICE:
for the Strategic Extension Campaign on
Integrated Weed Control
in the Muda Irrigation Scheme, Malaysia**

Dry seeding
WHY?
- To ensure two rice crops per year under water scarcity
- Double cropping can ensure continuous income to rice farmers
- Can cultivate rice even in drought situation
- To ensure income from rice cultivation even in drought situations

Selection of seeds
WHY?
- Reduce contamination by weed seeds
- Less reduction in rice yield
- Reduce weed problems
- Save costs in herbicide application
- Improve rice quality
- Increase rice yield
- Higher income from rice crop

FROM WHERE?
- Select seeds from plots without barnyard grass infestation
- Do not select seeds direcly from the combine harvester for planting
- Buy seeds which are of higher quality from Farmers' Association

→ **STEPS IN WEED CONTROL IN RICE FIELDS UNDER DRY SEEDING**

A. Land Preparation

1. Burn straw and rice stubbles

2. Rotovate land under dry condition

3. Rotovate land for second time to achieve better tilth

4. Level land with rear bucket attached to tractor

5. Construct in-field feeder channels (waterways) in rice fields to facilitate water distribution

6. Eradicate volunteer seedlings and weeds that germinate with paraquat at 2 litres/ha

7. Do not disturb the soil after paraquat application.
To disturb soil after paraquat application can encourage weed germination

LAND LEVELLING
WHY?
- Will facilitate water distribution in fields (water will be easily distributed to the whole plot)
- To reduce weed infestation
- To reduce use of herbicides
- Increase the effectiveness of herbicides
- Reduce cost in weed control
- Make weed control easier

PARAQUAT APPLICATION
WHY?
- Kill volunteer seedlings and weeds germinate before broadcasting
- Destroy sources of rice diseases and pests

CONSTRUCTION OF IN-FIELD FEEDER CHANNELS (WATERWAYS)

WHY?

- To facilitate uniform water distribution without using the water pump
- Cost saving
- Good water management reduces weed growth

HOW?

- In areas experiencing water delay, quarternary canals need to be constructed with majority agreement. Size of quaternary canal should be 12" wide and 12" deep along field levee (bunds). Distribute water to end of plot first
- In higher areas use the pump as a group activity for everybody's convenience
- The bund is important for water control. Size of bund must be 1' high and 1' wide and must be clean
- Field weir must be constructed and blocked with planks Farmers must visit fields everyday to check for efficient water management

B.   Seeds Broadcasting

8.   3-5 days after paraquat application, broadcast selected viable seeds at 70 kg/ha.

9.   Do not rotovate soil after broadcast

10.   Let in water into fields until soil is just wet

11.   Reseeding in vacant spots where germination is poor

12.   Maintaining standing water after seedling establishment

C. Herbicide Use

13. Spray STAM F-34 (Propanil) using fan-jet or floodjet nozzle. Rate of application is 2.5 kg a.i./ha. (a.i. = active ingredient) to control barnyard grass and Leptochloa grass at 2-3 leaves stage

or

Spary ARROSOLO (Propanil + Molinate) using polyjet nozzle at 3.0 kg a.i./ha. for barnyard grass anf Leptochloa grass at 2-3 leaves stage

14. 2-3 days after herbicide application, raise water level to 2-3 inches in rice field

15. Maintain water level in rice field

C A U T I O N

Do not use any insecticide
14 days before and after application
of STAM F-34 or ARROSOLO

16. Fill up empty spaces in the fields with healthy rice seedlings, and also pull out any barnyard grass and Leptochloa grass found in the field.

FIGURE 4-35

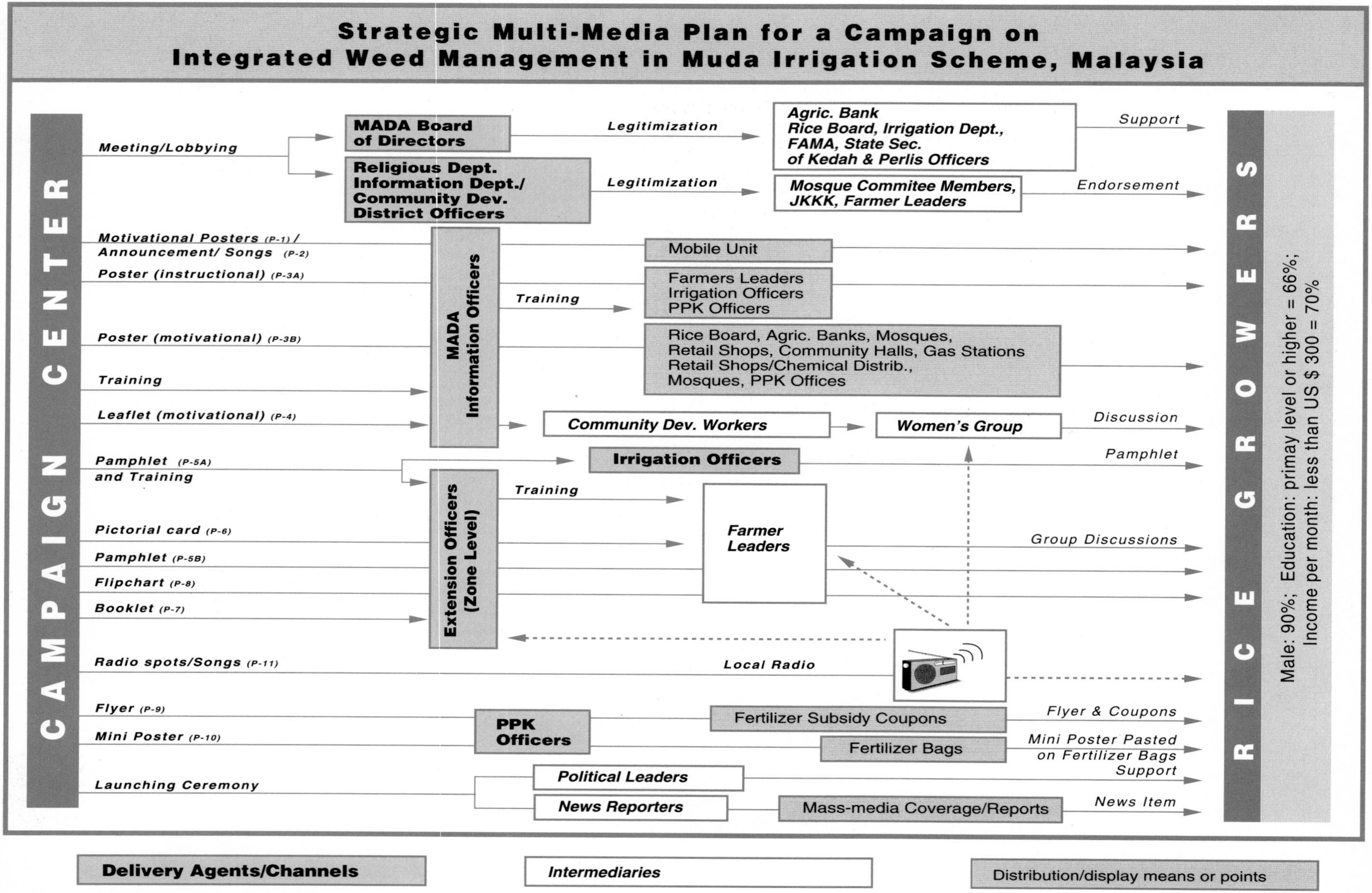

Strategic Multi-Media Plan for a Campaign on Integrated Weed Management in Muda Irrigation Scheme, Malaysia

RICE GROWERS

Male: 90%; Education: primay level or higher = 66%;
Income per month: less than US $ 300 = 70%

Support

Endorsement

Discussion

Pamphlet

Group Discussions

Flyer & Coupons

Mini Poster Pasted
on Fertilizer Bags
Support

News Item

Agric. Bank
Rice Board, Irrigation Dept.,
FAMA, State Sec.
of Kedah & Perlis Officers

Mosque Commitee Members,
JKKK, Farmer Leaders

Mobile Unit

Farmers Leaders
Irrigation Officers
PPK Officers

Rice Board, Agric. Banks, Mosques,
Retail Shops, Community Halls, Gas Stations
Retail Shops/Chemical Distrib.,
Mosques, PPK Offices

Women's Group

Fertilizer Subsidy Coupons

Fertilizer Bags

Mass-media Coverage/Reports

Legitimization

Legitimization

Training

Community Dev. Workers

Irrigation Officers

Farmer
Leaders

Local Radio

MADA Board
of Directors

Religious Dept.
Information Dept./
Community Dev.
District Officers

MADA
Information Officers

Extension Officers
(Zone Level)

Training

Training

PPK
Officers

Political Leaders

News Reporters

Meeting/Lobbying

Motivational Posters (P-1) /
Announcement/ Songs (P-2)
Poster (instructional) (P-3A)

Poster (motivational) (P-3B)

Training

Leaflet (motivational) (P-4)

Pamphlet (P-5A)
and Training

Pictorial card (P-6)

Pamphlet (P-5B)

Flipchart (P-8)

Booklet (P-7)

Radio spots/Songs (P-11)

Flyer (P-9)

Mini Poster (P-10)

Launching Ceremony

CAMPAIGN CENTER

Distribution/display means or points

Intermediaries

Delivery Agents/Channels

FIGURE 4-36

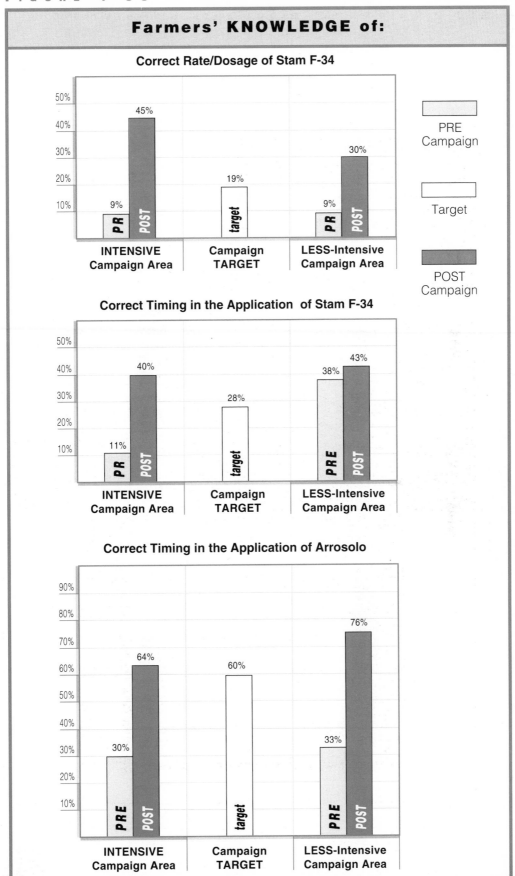

Farmers' KNOWLEDGE of:

Correct Rate/Dosage of Stam F-34

Correct Timing in the Application of Stam F-34

Correct Timing in the Application of Arrosolo

Source: Y.L. Khor and R. Mohamed, "The Information Recall and Impact Survey (IRIS) on the Strategic Extension Campaign on Integrated Weed Management in the Muda Irrigation Scheme, Malaysia", Penang, Malaysia, March 1990

FIGURE 4-37

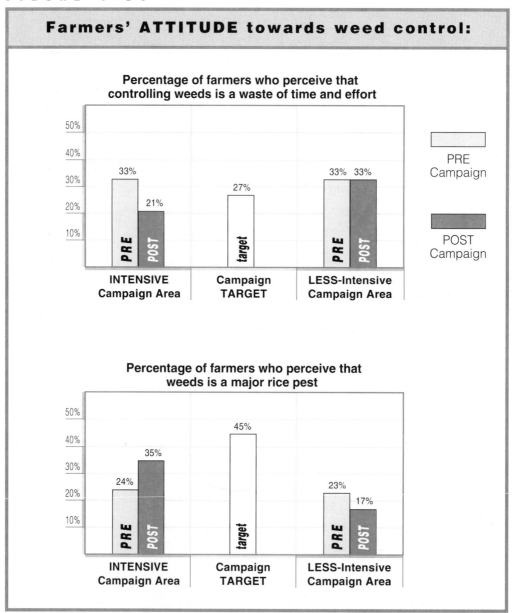

Farmers' ATTITUDE towards weed control:

Percentage of farmers who perceive that controlling weeds is a waste of time and effort

- INTENSIVE Campaign Area: PRE 33%, POST 21%
- Campaign TARGET: target 27%
- LESS-Intensive Campaign Area: PRE 33%, POST 33%

PRE Campaign
POST Campaign

Percentage of farmers who perceive that weeds is a major rice pest

- INTENSIVE Campaign Area: PRE 24%, POST 35%
- Campaign TARGET: target 45%
- LESS-Intensive Campaign Area: PRE 23%, POST 17%

Source: Y.L. Khor and R. Mohamed, "The Information Recall and Impact Survey (IRIS) on the Strategic Extension Campaign on Integrated Weed Management in the Muda Irrigation Scheme, Malaysia", Penang, Malaysia, March 1990

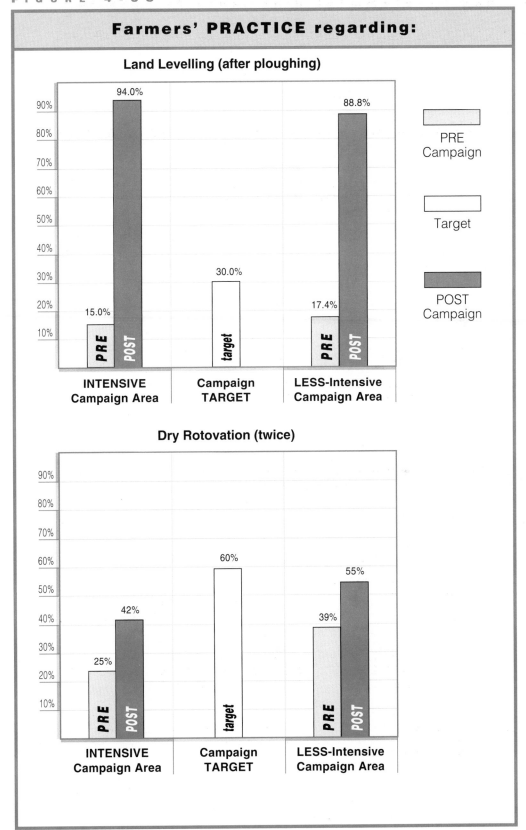

Farmers' PRACTICE regarding:

Land Levelling (after ploughing)

94.0% — INTENSIVE Campaign Area (POST)
15.0% — INTENSIVE Campaign Area (PRE)
30.0% — Campaign TARGET
17.4% — LESS-Intensive Campaign Area (PRE)
88.8% — LESS-Intensive Campaign Area (POST)

PRE Campaign
Target
POST Campaign

INTENSIVE Campaign Area
Campaign TARGET
LESS-Intensive Campaign Area

Dry Rotovation (twice)

25% — INTENSIVE Campaign Area (PRE)
42% — INTENSIVE Campaign Area (POST)
60% — Campaign TARGET
39% — LESS-Intensive Campaign Area (PRE)
55% — LESS-Intensive Campaign Area (POST)

INTENSIVE Campaign Area
Campaign TARGET
LESS-Intensive Campaign Area

Source: Y.L. Khor and R. Mohamed,   "The Information Recall and Impact Survey (IRIS) on the Strategic Extension Campaign on Integrated Weed Management in the Muda Irrigation Scheme, Malaysia", Penang, Malaysia, March 1990

FIGURE 4-39

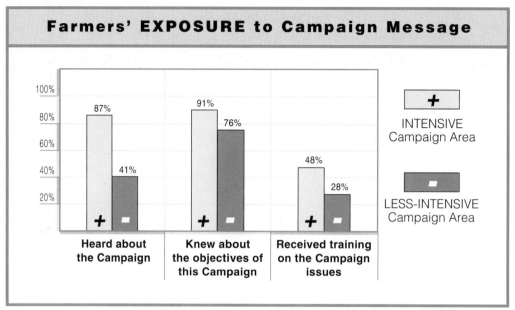

Source: Y.L. Khor and R. Mohamed, 1990

FIGURE 4-40

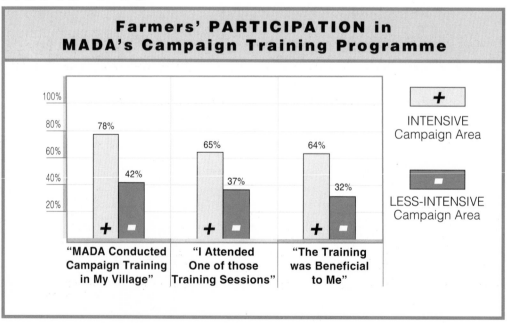

Source: Y.L. Khor and R. Mohamed, 1990

FIGURE 4-41

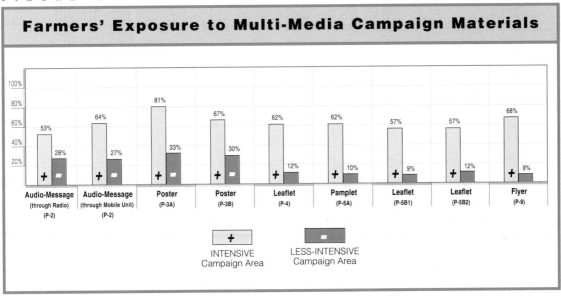

Source: Y.L. Khor and R. Mohamed, 1990

FIGURE 4-42

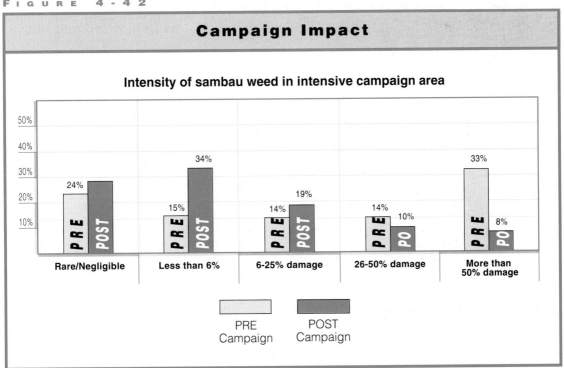

Source: Y.L. Khor and R. Mohamed, " The Information Recall and Impact Survey (IRIS) on the Strategic Extension Campaign on Integrated Weed Management in the MUDA Irrigation Scheme, Malaysia", Penang, Malaysia, March 1990.

FIGURE 4.43

Training Cost of the Integrated Weed Management Campaign in MUDA Irrigation Scheme, Kedah, Malaysia

EXPENDITURE	COST IN US $	PERCENTAGE
Honorarium/training allowance	4,285	38.9
Food & mileage claims	3,495	31.7
Office supplies, etc.	1,926	17.5
Equipment rental	1,310	11.9
Total	**11,016**	**100.0**

FIGURE 4.44

Breakdown of Expenses for Campaign Multi-Media Materials Development

EXPENDITURE	QUANTITY	COST IN M $	COST PER UNIT IN M $
Production of Logo	-	330	-
Production of drama and songs for radio and mobile unit	-	370	-
Instructional Posters	1,000	1,800	1.80
Motivational Posters	2,200	4,180	1.90
Motivational Leaflets	30,000	4,200	0.14
Instructional Pamphlets	35,000	4,900	0.14
Instructional Pamphlets (in Rumi)	20,000	4,200	0.21
Instructional Pamphlets (in Jawi)	15,000	3,300	0.22
Picture Cards	820	4,100	5.00
Flipcharts	519	11,418	22.00
Instructional Booklets	1,000	6,300	6.30
Total		**M $ 45,098 or US $ 16,703**	

US$ 1.00 = M$ 2.70 (based on exchange rate in 1989)

FIGURE 4-45

Cost and Benefit Analysis of Integrated Weed Management Campaign (in US $)

Acreage loss:	1988 =	4,938 hectares
	1989 =	2,661 hectares

Production gain after campaign =   2,277 hectares
Production estimates   1 hectares = 4.2 tons = $ 244.45

Financial gain after the campaign = 2277 x 4.2 x $ 224.45 = **$ 2,337,773**

Total savings = $ 2,337,773
Total campaign expenditure = $ 46.409

Cost/benefit ratio =   **1 : 50**

For each $ 1 invested, a return of $ 50 was gained.

Campaign Target   30,000 farm families

Estimated farm families
who adopted/practised   40% of target audience = 12,000 farm families
campaign recommendations

Average economic benefit   US$ 2,337,773 : 12,000 = **US$ 195. -**
per farm family who adopted
campaign recommendations

FIGURE 4-46

Total Expenditure for the Integrated Weed Management Campaign in Muda Irrigation Scheme, Kedah, Malaysia

EXPENDITURE	COST IN US $	PERCENTAGE
SEC Training (5 workshops)	11,016	23.7
Design and production of campaign materials (print and broadcast)	16,703	35.9
Campaign launching	1,926	4.2
Training and distribution of campaign materials	3,444	7.5
Research and evaluation	12,570	27.1
Field demonstration	750	1.6
Total	**46,409**	**100.0**

FIGURE 4.47

Yield Performance and Herbicide Usage (1987-1993) in the MUDA Agricultural Development Authority (MADA) BEFORE and AFTER the Strategic Extension Campaign (SEC) on Integrated Weed Management (IWM)

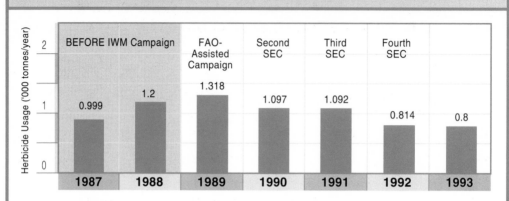

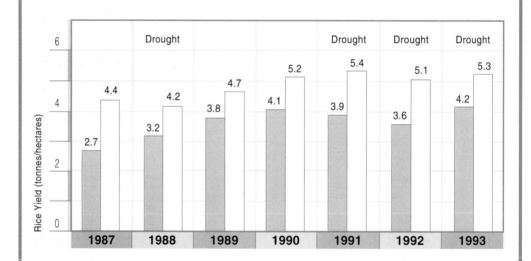

Note:
Provision of agricultural supplies/ imputs by MADA was not changed during this period.

Source:
Nai-Kin Ho (1994),
"Integrated Weed Management of Rice in Malaysia: Some
Aspects of the MUDA Irrigation Scheme's Approach and Experience".

4.5.
Zambia: Assisting Small Farmers through Maize Production Campaign*

During 1990-1992, Strategic Extension Campaign (SEC) activities on maize production were undertaken in Zambia by an FAO/UNDP project ZAM/88/021 "Strengthening Agricultural Extension Services of the Department of Agriculture, Southern Province". The project which was operational from June 1989 to December 1992 provided technical support in improving agricultural extension methods and services in order to contribute to the development of agricultural sector in Zambia's New Economic Recovery Programme. With a budget of US$ 1.2 million (for 3.5 years), the project's target beneficiaries were 80,000 small farm families in the Southern Province of Zambia. One of the major project activities was the planning and implementation of the Strategic Extension Campaign (SEC) on Maize Production, including training of the Dept. of Agriculture staff in SEC methods and techniques.

4.5.1. *Staff Development*

Provincial and District Crop Husbandry Specialists of the Department of Agriculture in Southern Province were given training on effective methodology of field inquiry and reconnaissance surveys, with a view to identify farmers' main problems and constraints and consequently to provide solution, through development of area specific problem-solving oriented agronomic, livestock, post-harvest losses practices and messages. The instrument for the farmers' KAP (Knowledge, Attitude, and Practice)/baseline survey on maize production was completed in November 1990.

---

* This section is based heavily on the Terminal Report of project ZAM/88/021, Rome: FAO/United Nations, 1993

Extension personnel were trained in KAP survey implementation techniques in February 1991, and they conducted the KAP/baseline survey on maize production through personal interviews and focus group interviews with farmers from March to July 1991.

The findings from the KAP/baseline survey were analyzed and utilized in Strategic Extension Campaign (SEC) workshops in October and November 1991 and again in February and September 1992. These workshops aimed at planning and developing a campaign strategy and designing prototype messages and multi-media materials based on the identified problems for the SEC on Maize Production in Southern Province. Additional aims of the workshop were: to provide extension workers with a theoretical and practical understanding and conceptual framework of strategic planning and management; to demonstrate how to formulate objectives and strategies for extension programme planning, including problem identification, objective formulation, strategy development and audience segmentation; to impart skills and techniques on multi-media campaign planning, strategy development, message design, and pretesting of prototype multi-media materials.

At least eight provincial and district Crop Husbandry Specialists, 55 Block Supervisors, 48 Subject-Matter Specialists, and 202 Camp Officers, were trained on various SEC activities in Southern Province. A wide-variety of multi-media materials covering 17 different types of messages were developed, pretested, and reproduced in large quantities in 1991. The Maize Production campaign was conducted in 1991 - 1992, and later in-mid 1992 new campaign topics, such as livestock (i.e., cattle), sorghum, sunflower and post-harvest practices were also introduced.

4.5.2.   *Applying SEC for Extension Programme Planning*

Using SEC principles and concepts, the project developed a strategic extension programme cycle, focusing on the development of problem-oriented, area-specific agronomic, livestock, and post-harvest practices. The approach appears to be highly suitable for rain-fed, drought-prone, smallholder problems at hand, as it emphasizes a "bottom-up" approach to extension strategy, programme planning, and message development rather than a "top-down" approach determined by external research and development trends. The project also concentrated more on strategic extension programme development and in-service training using experiential learning principles as applied during SEC workshops. Extension staff's improved competency was evidenced in their closer and more effective relationship with small farmers. SEC field-based activities had served as an important and valuable function to stimulate extension workers' interests in new ideas and innovative approaches in working with farmers.

One of the results of the SEC activities that has important implications on future extension programme planning was that due to the information and education demand-creation effects of campaign messages, farmers are requesting for more frequent and specific extension and training services from extension workers. Once stimulated or made aware of the campaign issues or messages through multi-media channels or materials, farmers in Southern Province sought addition information from neighbours, friends, extension workers, and/or progressive farmers in the area. Farmers and other interested groups had approached agricultural camps, blocks, district and provincial offices requesting for more copies of extension support materials and more messages to be developed. Fortunately, there were enough copies of the campaign media materials in local languages and their distribution was also timely. The experience of the project has been that the demand from farmers and extension workers for more extension and training activities was greater than expected, as a result of the campaign activities.

<hr>

4.5.3. *Project Evaluation*

An in-depth evaluation of ZAM/88/021 project was carried out in March 1992. The independent evaluation mission led by UNDP found that the project has been successful in its implementation. The project was highly commended for the strategic extension campaign (SEC) methodologies used, high implementation rate (average 98 percent), monitoring and evaluation activities, and

over-all technical and operational soundness. In its conclusion the evaluation mission highly recommended that the project be expanded and donor funding continued into the second phase.

Excerpt from the Report of the Evaluation Mission - project ZAM/88/021 by W.M. Rivera, A.M. Bannaga, and J. Phiri, New York : UNDP, 1992

"The evaluation mission found that the project adopted and advanced a program development rather than a top-down technology delivery approach to extension, utilizing a 'strategic extension campaign' system developed by the FAO. This approach emphasizes the development of problem-oriented, area-specific agronomic, livestock, and ...post harvest practices. The team found this approach highly suitable for the rain-fed, drought-prone, smallholder problems at hand, as it emphasizes a so-called "bottom-up approach to extension message formulation... The mission team was favorably impressed by the project strategy and its methodology, including its management systems for training, monitoring, and evaluation."

"The project's utilization of KAP (Knowledge, Attitude, Practice) baseline surveys indicates that such surveys are imperative for the development of a bottom-up extension strategy and crucial in identifying problems that are both relevant to extension and can be tackled by extension staff. Household studies are not sufficient for developing extension strategies, although they may be useful. Extension program leaders need to know first what local famers know about farming and what their attitudes and practices are with respect to their farming systems.

The last Tri-Partite Review meeting (i.e., Government of Zambia, UNDP, and FAO) held in October 1992 gave a sound assessment of the project. The Government of Zambia regarded the ZAM/88/021 project as "one of the best agricultural extension projects, if not the best in Zambia so far". UNDP regarded the project as "an uplifting and successful technical cooperation project".

During UNDP's "Putting People in Centre of Development" meeting held in Lusaka on 18th, August 1992, the ZAM/88/021 project was selected as the best project among all UNDP funded, and also among the projects implemented by various UN agencies and non-government organizations (e.g. Africa 2000 Network) in Zambia. Project 2AM/88/021 was considered to have made relevant impact and offered an excellent programme in human resources development. Hence, its programme and achievements were recommended to UNDP Headquarters to appear in the UNDP 1993 Human Development Report publication which is widely circulated.

Planning Strategic Extension Campaign in Zambia

Zambia: Maize Production Campaign

Developing Multi-Media Campaign Messages and Materials in Zambia

Field Extension and Training Activities during a Strategic Extension Campaign on Maize Production in Zambia

4.6.
Applying SEC
in Population Education
through Agriculture
Extension Programmes

Basic concepts and principles of Strategic Extension Campaign (SEC) had also been used and adapted in the efforts to integrate population issues/concerns into the agricultural extension activities in several FAO member countries. With funding from the United Nations Population Fund (UNFPA), FAO/ESHE executed and operated an inter-regional project INT/88/P28 "Strategic Integration of Population Education into the Agricultural Extension Services (PEDAEX)". This PEDAEX project with a total budget of US$ 1.25 million was in operation from September 1986 to December 1991. Its main objective was to develop strategies, methodologies, and conduct pilot activities to integrate systematically population education into national agricultural extension services in selected countries.

4.6.1. *PEDAEX Results*

The project was successfully completed in December 1991 with pilot activities conducted in eight countries : Honduras, Jamaica, Kenya, Malawi, Morocco, Tunisia, Rwanda and Thailand. Important output-oriented activities, following a PEDAEX operational process based on an SEC conceptual framework, were undertaken in the eight countries. These activities included among others: PEDAEX planning and strategy development workshop, Knowledge, Attitude, and Practice (KAP) baseline surveys, establishment of PEDAEX task forces/steering committees, workshops on PEDAEX strategy planning, message design & multi-media materials development, extension workers training on the use of specifically-designed PEDAEX extension/training support materials, and management of PEDAEX field implementation. Process evaluation and summative evaluation studies of PEDAEX activities were also conducted in the eight countries.

Evaluation results and lessons learned from PEDAEX project experiences were shared in an international workshop on "PEDAEX Experience Sharing and Results Demonstration" held in Meknès, Morocco in November 1991. A summary of these main activities can be seen in Figure 4-47.

The intended project objectives have been fully achieved. Figure 4-48 shows a summary of Project's Performance Indicators and Outputs. The PEDAEX strategy and methodology, as well as extension/training support materials developed and tested by the project appeared to be appropriate, effective and useful. In some countries, it was also felt that PEDAEX activities helped to revitalize, and enhance the effectiveness of, the extension service. Understanding on how population-related aspects can positively or negatively affect farm families' agricultural productivity and well-being in particular, and food security in general, is now considered a critical factor in facilitating the effectiveness of extension and training activities for sustainable and environmentally-sound agricultural development.

Follow-up PEDAEX activities are underway, including plans for wider-scale PEDAEX replications. Due to the participatory approach in developing PEDAEX strategy, methodology, and extension/training support materials, and the emphasis in training of local trainers and extension officers, PEDAEX activities also seem more likely to be sustainable and institutionalized. More specifically, during the project period, the following results were accomplished :

- International Workshop on "Planning and Strategy of Integrating Population Education into Agricultural Extension Services" held in Rome in May 1987 with participants from 10 countries.

- PEDAEX pilot activity project proposals for 8 countries developed and approved for funding by UNFPA.

- PEDAEX task forces/steering committees established in 8 countries.

- Baseline/KAP studies conducted in 8 countries.

- Workshops on "PEDAEX Strategy Planning, Message Design and Extension/ Training Materials Development" held in 8 countries for 191 participants.

- Various specifically-designed prototype PEDAEX extension/training support materials pretested, revised and reproduced in 8 countries.

- A total of 286 extension officers in 8 countries trained in management of PEDAEX field implementation and use of multi-media PEDAEX extension/ training support materials.

- PEDAEX field implementation is now carried out in 8 countries, and as of Dec. 1991, it was reported that an estimated total of at least 102,000 farmers and/or their family members had been reached.

- Three travelling seminars/process evaluation activities conducted in four regions (Asia, Africa, the Near East and Latin America & Caribbean).

- Summative evaluation studies of PEDAEX activities conducted in 8 countries.

- International Workshop on "PEDAEX Experience Sharing and Results Demonstration" held in Morocco in Nov. 1991 with participants from 16 countries.

Logos of SEC Activities on Population Education through Agricultural Extension Programmes

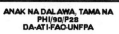

FIGURE 4-47

FAO/UNFPA Project INT/88/P28 on:
"Strategic Integration of Population Education into the Agricultural Extension Services (PEDAEX)"
PEDAEX Activities Implementation

IMPORTANT PEDAEX ACTIVITIES / COUNTRY	PLANNING WORKSHOP	PILOT ACT. PREPARATION	KAP/BASELINE SURVEY	STRATEGY PLANNING, MESSAGE DESIGN, MATERIALS DEV. WORKSHOP	PRETESTING, REVISIONS & PRODUCTION OF MATERIALS	TRAINING OF EXTENSION STAFF	PEDAEX FIELD IMPLEMENTATION (started in:)	PROCESS EVALUATION/ TRAVELLING SEMINAR	SUMMATIVE EVALUATION
ASIA									
THAILAND	MAY 1987	MAR. 1989	MAR. 1990	APR. 1990	JUN. 1990	AUG. 1990	AUG. 1990	MAR. 1991	OCT. 1991
AFRICA									
MALAWI	MAY 1987	APR. 1988	DEC. 1988	SEP. 1989	JAN. 1991	MAR. 1991	MAR. 1991	MAR. 1991	OCT. 1991
KENYA	MAY 1987	DEC. 1988	MAR. 1990	JUL. 1990	NOV. 1990	JAN. 1991	FEB. 1991	MAR. 1991	OCT. 1991
RWANDA	MAY 1987	JUN. 1988	NOV. 1990	JUL. 1991	OCT. 1991	SEP. 1991	SEP. 1991	JUN. 1991	MAR. 1992
NEAR EAST									
MOROCCO	MAY 1987	JUL. 1988	NOV. 1989	MAR. 1990	DEC. 1990	JAN. 1991	FEB. 1991	JUN. 1991	OCT. 1991
TUNISIA	MAY 1987	JUN. 1988	MAY 1989	DEC. 1989	SEP. 1990	NOV. 1990	NOV. 1990	JUN. 1991	OCT. 1991
LAT. AMERICA/ CARIBBEAN									
HONDURAS	MAY 1987	JUL. 1988	DEC. 1989	AUG. 1990	JUN. 1991	APR. 1991	APR. 1991	JUN. 1991	OCT. 1991
JAMAICA	MAY 1987	JAN. 1989	JUL. 1990	OCT. 1990	FEB. 1991	MAR. 1991	MAR. 1991	JUN. 1991	OCT. 1991

RA/ESHE: updated March 1992

FIGURE 4-48

FAO/UNFPA Project INT/88/P28 on "Strategic Integration of Population Education into the Agricultural Extension Services (PEDAEX)"
PEDAEX PERFORMANCE INDICATORS AND OUTPUTS

COUNTRY	THAILAND	KENYA	MALAWI	RWANDA	MOROCCO	TUNISIA	JAMAICA	HONDURAS	TOTAL
Pilot Activity Project No.:	THA/89/P04	KEN/89/P01	MLW/88/P04	RWA/88/P01	MOR/88/P07	TUN/88/P02	JAM/89/P02	HON/88/P05	
BUDGET									
Cost of Pilot Activity Projects	$ 27,414	$ 28,169	$ 40,100	$ 25,850	$ 42,796	$ 38,899	$ 25,786	$ 25,350	$ 254,364
Inter-Regional Project Cost (INT/88/P28)									$ 1,003,236
Total Cost									$ 1,257,600
TARGET AUDIENCE (TOTAL NO.)	2,970	14,304	13,886	240,000	150,000	10,000	1,230	23,000	455,390
Small Farmers	Yes	Yes	Yes	Yes	Yes	Yes	Yes	Yes	
Male Field Extension Workers	Yes	Yes	Yes	Yes	Yes	Yes	Yes	Yes	
Female Ext. Workers	Yes	Yes	Yes	Yes	Yes	Yes	Yes	Yes	
Female Farmers	Yes	Yes	Yes	Yes	Yes	Yes	Yes	Yes	
Rural Youth	Yes	Yes	Yes	Yes	Yes	Yes	Yes	Yes	
COORDINATING TASK GROUP Consisting of Min. of Agriculture's Extension Dep. and the following Ministries, Departments, and/or Agencies:	Dept. of Tech. & Eco. Coop. (DTEC), Nat. Eco.& Soc. Dev. Brd./NESDB Kasetsart Univ.	Health, Educ., Nat. Comm. for Population and Dev. (NCPD)	Health, Comm. Serv., Demography Unit of Univ. of Malawi, Office of the President & Cabinet	Education, Population (ONAPO)	Public Health, Education, Youth & Sports, Social Affairs, Interior	Education, Population (ONFP)	Nat. Fam. Plan. Brd., Health, Edu., Jam.Agri. Society, Planning Inst. of Jam.	Health, Education, Welfare, National Univ.	
KAP/BASELINE SURVEY									
Conducted by:	Contractor	Contractor	Contractor	Contractor	Min. of Agri.	Contractor	Contractor	Contractor	
Survey Sample Size (respondents)	300	184	156	230	362	200	120	200	
Focus Group Interviews (FGI)	No	Yes	Yes	Yse	Yes	Yes	Yes	Yes	
Number of Groups for FGI	-	4	4	5	4	3	3	2	
Total FGI Respondents	-	46	76	75	47	32	55	11	
Duration of Study (in months)	3 months	7 months	4 months	18 months*	8 months	4 months	8 months	4 months	
Cost (in US $)	$ 2,500	$ 4,000	$ 4,700	$ 4,000	$ 3,000	$ 2,500	$ 5,536	$ 3,000	$ 29,236
STAFF TRAINING/WORKSHOP ON:	No. of Participants	No. of Participants	No. of Participants	No. of Participants	No. of Participants	No. of Participants	No. of Participants	No. of Participants	Total
■ Seminar for Policymakers/Planners	27		7			9		9	52
■ Strategy Planning, Message Design, & Multi-media Materials Dev. and Pretesting (Training of Trainers)	42	14	28	22	20	23	19	23	191
■ Training of Extension Workers and/ or Subjectt-Matter Specialists (SMSs)	32	75	23	54	28	15	28	31	286
Total Training Participants	101	89	58	76	48	47	47	63	529

RAJ/ESHE: updated May 2, 1992
* Delays due to political unrest/security situation in 1991

FIGURE 4-48 (CONTINUED)

FAO/UNFPA Project INT/88/P28 on "Strategic Integration of Population Education into the Agricultural Extension Services (PEDAEX)"
PEDAEX PERFORMANCE INDICATORS AND OUTPUTS (continued)

	THAILAND		KENYA		MALAWI		RWANDA		MOROCCO		TUNISIA		JAMAICA		HONDURAS		TOTAL
PEDAEX EXTENSION/TRAINING **MATERIALS PRODUCED**	Type(s)	Total Quan.	Type(s)	Total Quan.	Type(s)	Total Quan.	Type(s)	Total Quan.	Type(s)	Total Quan.	Type(s)	Total Quan.	Type(s)	Total Quan.	Type(s)	Total Quan.	Total Quant.
Posters	2	2,000	5	5,000	3	900	5	2,500	7	6,500	2	1,000	2	400	2	1,000	19,300
Handbook/manual	1	1,000	1	500	1	100	1	100	1	500	1	100	1	200	1	240	2,740
Leaflets/Pamphlets	3	3,600	2	4,000	1	1,000	4	2,000	2	5,000	3	1,500	3	4,500	2	3,000	24,600
Flipcharts	1	200	2	1,000	1	100	1	100			1	100	1	100	2	200	1,800
Stickers	2	400	5	5,000			3	1,500	1	5,000	1	500	1	500			12,900
Audio-Cassette Programmes									1	50					2	50	100
Radio Programmes (total in minutes)							6	93							52	1,300	1,393
Bill Boards													1	1	1	10	11
T-Shirts																	0
Video Cassette	1	10							1	40					1	15	65
Slide-Sound Sets					1	126			1	1,250					2	300	1,676
Others (pls. describe): Overhead Transparency Sets	1	10															10
Calendars					1	300											300
Cost of Materials (incl. design, pretest & reproduction)		$6,500		$9,347		$5,235		$11,500		$29,640		$11,000		$12,342		$11,500	$97,064
FIELD IMPLEMENTATION/ **OUTREACH ACTIVITIES**	No.	Total	No.	Total	No.	Total	No.	Total	No.	Total	No.	Total	No.	Total	No.	Total	Total
Farmer Leaders Training	6		11		5								12		9		
Est. Number of Leaders Trained		35		270		265		480		8,166				480		60	9,756
Farmers Group meetings	36		19		558		212		217				15		40		
Est. Number of Farmers Reached		1,200		5,960		13,886		55,687		9,879		1,844		900		568	89,924
Others (pls. describe): School Lectures															31		
Est. Number of Students Reached																2,480	2,480
Est. No. of Persons Trained	1,235		6,230		14,151		56,167		18,045		1,844		1,380		3,108		102,160
SUMMATIVE EVALUATION																	Total Cost
Conducted by:	Contractor		Contractor		Contractor		Contractor		Contractor		Contractor		Contractor		Contractor		
Survey Sample Size (respondents)	69		161		132		120		240		200		202		200		
Focus Group Interviews (FGI)	Yes		Yes		Yes		No		Yes		Yes		Yes		Yes		
Number of Groups for FGI	4		4		4		-		5		3		2		2		
Total FGI Respondents	40		73		132		-		55		30		51		11		
Duration of Study (in months)	3 months		3 months		3 months		3 months		3 months		3 months		3 months		2.5 months		
Cost (in US $)	$4,000		$4,400		$4,000		$4,000		$3,000		$2,500		$4,500		$3,000		$29,400

RA/ESHE: updated May 2, 1992

4.6.2.   *Future Directions for PEDAEX Replications*

One of the most important conclusions and recommendations based on the project's results and experience is the need to provide a systematic and quality-controlled PEDAEX training programme following the SEC training approach. Such a training programme with specifically-tailored PEDAEX curricula, syllabi, lesson plans, teaching aids/learning materials, etc., which can be flexibly adapted or modified for specific country situations, problems and needs, would facilitate and contribute greatly in developing PEDAEX know-how among local agricultural extension and training personnel.

Training in PEDAEX management should also be encouraged because an important element in the PEDAEX concept is the need for inter-ministerial/departmental collaboration. Proper PEDAEX planning skills are especially needed in view of the need of joint field programming activities among concerned agencies/institutions. The demand-creation effects generated through population education by agricultural extension workers must be well-synchronized, through effective referral services, with other relevant field inputs/supply delivery agencies (i.e., health, family planning, environment, women affairs, etc.) to avoid a frustration of rising expectation situation among rural families.

4.6.3.   *Lessons Learned from PEDAEX Activities*

Based on the results of FAO/UNFPA project INT/88/P28 on "Strategic Integration of Population Education into the Agricultural Extension Services (PEDAEX)", as well as the findings of the process and summative evaluation studies conducted for the project's pilot activities in the eight participating countries (Kenya, Malawi, Rwanda, Tunisia, Morocco, Honduras, Jamaica, and Thailand), the following lessons have been learned from the PEDAEX experiences :

- Successful PEDAEX activities require functioning and well-organized agricultural extension services which are fully operational at the field-level, in order to integrate effectively population education into regular and institutionalized agricultural extension & training programmes.

● Essential policy support from within the Ministry of Agriculture must be mobilized by providing important policy/decision- makers with clear and explicit rationale for PEDAEX, including its comparative advantages, especially on the concrete and practical benefits to both the agricultural extension service and its main clientele, —small farmers, rural women and youth.

● Inter-ministerial/departmental/agency collaboration or coordination, at least in the form of a PEDAEX task-force or steering-committee, is needed for obtaining advice or inputs on appropriate and relevant technical subject matters related to PEDAEX. Such a collaborative and coordinated effort can also help to synchronize field implementation programming and activities, especially with agencies responsible for providing rural health and family planning inputs/service delivery.

● PEDAEX task-force meetings, if held regularly, can be very effective especially for identifying priority population education contents/messages, and for joint programming/planning and monitoring of field implementation activities.

● Problem-solving oriented PEDAEX activities can be most effective, especially if problem identification and/or needs assessment exercises are conducted, using a participatory approach involving the target audience, through a specifically designed baseline or Knowledge, Attitude & Practice (KAP) survey, including appropriate focus group interviews (FGI).

● KAP surveys' and focus group interviews' results are most useful for PEDAEX's objective formulation, strategy development and planning tasks, if the survey focuses more on relevant issues related to the inter-relationships of population issues and agricultural development, or on population-related factors that will positively or negatively affect agricultural production or productivity at the farmers' level, instead only on population/family planning issues.

- The implementation of PEDAEX activities is more likely to succeed if undertaken in a systems-approach based on the suggested PEDAEX conceptual framework which identifies its basic processes, elements/components and operational steps, rather than conducting only parts of its elements/components or activities on an ad-hoc basis.

- The involvement of the Ministry of Agriculture's subject-matter specialists (SMSs), in addition to its extension planners, trainers and field workers, is very much needed. Such involvement and collaboration are especially necessary in the important and difficult tasks of properly matching, integrating and packaging relevant population issues/contents with priority agricultural production/development messages, for dissemination through agricultural training and extension services.

- Due to the complexity and technical nature of the inter-relationship between population, agriculture, and also environment, PEDAEX messages cannot rely mainly on general, simplified, and slogan-oriented information. PEDAEX issues need to be communicated to, and understood by, the target audience. Such PEDAEX messages could be internalized better and easier by the target audience if they are specific, issue or problem-solving oriented, and locally-relevant.

- As the objective of PEDAEX is not limited to awareness creation on population-related issues, PEDAEX message dissemination tasks cannot be done solely through mass communication or mediated-communication channels. To facilitate proper understanding of the rather complicated nature of PEDAEX messages, the use of an educational approach, rather than a communication approach, is preferred. Such an educational approach can be effectively implemented through small group meetings or training to allow for more complex and comprehensive PEDAEX information/messages to be shared, discussed, analysed, and evaluated in order to enhance the target audience's learning process.

- The effectiveness in developing appropriate or relevant PEDAEX messages and population educational strategies depends largely on the quality of PEDAEX training programmes for extension planners, trainers, and field-workers. The use of appropriate and well-tested curricula, instructor's manual, training aids and learning materials can significantly improve PEDAEX training performance and results. Specifically designed and high-quality PEDAEX training materials are also critically needed to facilitate PEDAEX replications within a country or in other countries.

- Agricultural extension workers and trainers can undertake PEDAEX activities more effectively if they focus mainly on the informational, motivational, and educational aspects of population issues, and provide only referral services on contraceptive methods or family planning related issues to other competent and concerned agencies and/or their field workers.

- Sustainability and institutionalization of PEDAEX activities can be facilitated by employing a participatory planning and training approach involving local agricultural extension planners, trainers, field workers and farmer leaders in undertaking PEDAEX problems/needs identification, objective formulation, strategy development, message design and extension/training support materials development and testing, as well as field implementation, management and monitoring activities.

- Specifically designed and field-tested PEDAEX multi-media extension and training support materials, based on problems identified through KAP survey and the planned strategy, are critical for effective PEDAEX field implementation. Such PEDAEX materials, especially when these were attractively designed, developed and tested by the potential users themselves, do not only serve as essential aids to facilitate a high quality-controlled extension/training service delivery. The use of such PEDAEX multi-media materials by extension workers and trainers, also boosts their morale and enthusiasm, and increases their professional credibility and confidence. Although a considerable proportion of PEDAEX operating cost is required for the development, testing, and production of multi-media extension/training support materials, such an expenditure is well-justified in terms of its cost-effectiveness.

- A realistic and gradual/phased-approach to planning and management of PEDAEX field implementation, including PEDAEX staff assignments and budget allocations, is necessary in introducing and demonstrating PEDAEX as a viable and desirable programme which should become a part of a national agricultural extension service.

- PEDAEX activities can benefit significantly from the results of process/formative evaluation and summative evaluation studies, especially for making necessary modifications and/or improvements. Such evaluation studies as an integral part of the PEDAEX activities also proved to be useful for encouraging wider PEDAEX replications. Such replications can be further facilitated by a PEDAEX process documentation which provides important descriptions and/or analyses of major operational steps, results and experiences of PEDAEX activities.

PEDAEX Results Demonstration and Experience Sharing Workshop in Morocco

5. LESSONS LEARNED:
Usefulness of SEC for Improving Extension System and Programmes

       This Chapter will not discuss the lessons learned on how to plan, implement, monitor and evaluate SEC activities. To a large extent, such lessons have been drawn by Adhikarya with Posamentier (1987) from the experiences of the Bangladesh Rat Control Campaigns of 1983 and 1984. Instead, this Chapter will discuss the macro-level implications from the SEC experiences for strengthening and improving public-service agricultural extension systems, programmes and activities. It was pointed out earlier in Section 1.2 that SEC is not a substitute for an agricultural extension system or programme. SEC is only one of the non-formal education methods which should be an integral part of an agricultural extension system and/or programme. Experiences from various SEC applications have generated some important and useful lessons for increasing further the effectiveness and efficiency of agricultural extension system and programmes by applying some or all of the SEC elements and/or principles.

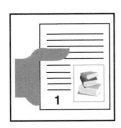

5.1.
SEC Enhances the Agricultural Extension Planning Process

       Strategic Extension Campaign (SEC) method places premium importance on a systematic procedure of assessing the felt needs of target beneficiaries (e.g., small farmers, rural women, youth, etc.) as well as the intermediaries (e.g. extension workers, trainers, subject-matter specialists, etc.), and in identifying their perceived problems or issues which might be the reasons for non-adoption or improper practice of a certain recommended technology. SEC experiences have indicated that the important principle of developing specific and precise extension objectives based on felt needs and perceived problems of the target audience (i.e., beneficiaries and concerned persons involved in agricultural exten-sion/outreach activities) is fundamental to ensuring the relevance and appropriate-ness of a strategically-planned and participatory-approach agricultural extension programme.

SEC activities have also shown that many extension plan-ners/managers and trainers who had been trained in SEC workshops, especially on the Use of Farmers' Knowledge, Attitude, and Practice (KAP) Survey, and on Strategy Planning, Message Design and Multi-Media Materials Development, have applied strategic planning principles. They have commissioned or conducted KAP/baseline surveys, including Focus Group Interviews (FGI) and used the quan-titative data and qualitative information as inputs for routine extension programme planning, monitoring and evaluation, thus beyond extension campaign purposes. Extension programmes and activities, as a result of such strategic planning and sys-tems-approach orientations, are thus more focused, needs-based, problem-solving, and have specific objectives and tailored-messages specifically designed for seg-mented target groups. Previous extension orientation seemed to be more of an ad-hoc and sporadic technology dissemination approach, with general objectives and sometimes vague or irrelevant messages.

Observations have also been made on the "strategic" decision making processes applied by those who have been involved in SEC planning, im-plementation, and management activities. Such strategic decisions are most clearly demonstrated in undertaking specific tasks, such as, prioritizing extension problems to solve, formulating specific and measurable extension objectives, determining relevant and appropriate extension messages and training contents, selecting the cost-effective combination of multi-media support materials to develop, pretest, and produce, management planning for extension/outreach activities, etc. The con-scious, planned, and systematic efforts reflected as a result of such strategic think-ing and decisions, can significantly help increase the cost-effectiveness and efficien-cy of agricultural extension programme performance as reported by many SEC programmes. There is clear evidence that some or all SEC elements are applicable and useful for improving further the performance and quality of regular agricultural extension system and programmes.

5.2.
SEC Builds Cadres of Extension Programme Planners and Trainers

In SEC activities, human resources development through staff training on the SEC methodology (see Figs. 2.2 and 2.3) is a very crucial element. It should be considered as a good "insurance policy" to ensure sustainability and institutionalization of the application or utilization of the SEC know-how in support of agricultural extension programme implementation. Through various practical hands-on, skills-development oriented, and experiential-learning method SEC workshops, a core-group of agricultural extension and training staff in a given country are trained to serve as SEC Master Trainers and Master Planners.

SEC experiences have shown that when there is an adequate number of SEC trained personnel in a given agricultural extension and/or training institution, in the process of SEC replications, they can effectively serve as multiplier agents in developing other SEC trained resource persons (see Section 4.2.4 and Fig. 4-19). Furthermore, some SEC Master Trainers do not only serve as resource persons in their respective countries, but they also proved to be effective in assisting SEC programme replications in other countries as shown in Fig. 3.1.

An important lesson learned from SEC experiences in different countries regarding its training strategy is that it does not create an unnecessary "dependency" on foreign resource persons and/or consultants, after the initial round of training of trainers activities have been completed. It seems possible to transfer SEC methodology through training of local extension planners and trainers, without having to perpetuate the need for external technical assistance after one or two SEC programmes have been implemented successfully.

5.3.
SEC Helps in Improving Extension Linkage with Research

In any SEC programme, a new or existing agricultural technology package must be identified and selected as the core contents for the development of extension messages. With the assistance of appropriate agricultural research staff or subject matter specialists, the recommended technology package must be validated, and the essential and/or critical elements of the technology must be specified. Without such specific sets of information, the level of knowledge, attitude, and practice (KAP) of farmers regarding the technology can not be determined through a KAP/baseline survey. SEC experience showed the importance, and useful results, of the collaboration between research personnel and extension staff in planning a KAP survey, and in formulating and prioritizing extension objectives based on the survey data.

SEC activities have also demonstrated that even when no "new" agricultural technology or innovation is available, extension services to most farmers on "existing" technologies and/or know-how are necessary and needed. Through a KAP survey, the specific status of a given technology adoption or practice among farmers, and more importantly, the reasons for non-adoption, and the degree of inappropriate practices, of such a technology, can be determined and analyzed. Such data, information and analysis are important inputs for research staff to improve the technology to make it more acceptable for proper adoption by farmers. Thus, the SEC's participatory process by facilitating a two-way communication between research-extension-farmers helps in setting the research agenda by providing "feed-forward" information regarding research needs as perceived by farmers. Such an agriculture research "agenda-setting" function seems to be performed effectively by SEC activities, especially regarding indigenous knowledge system or existing/known cultural practices, which have been practiced by some but not the majority of farmers.

5.4.
SEC is Needed Most by
Small, Resource-Poor
Farmers

It has been widely reported by many diffusion of innovation studies (Rogers, 1981) that "new technology" or innovation (incl. agricultural technology), as long as it shows tangible benefits, or an improvement over the old practice, such a technology will transfer or spread fairly easily, even without the help of an extension service. However, in such a case, the higher SES (socio-economic-status) group members of the community are those who are more likely to practice or adopt the recommended new technology (Rogers and Adhikarya, 1979). For the lower SES group members, the process of adopting and practicing properly the recommended technology is more than just rational decision-making based on risk analysis, economic benefits or other incentives/rewards considerations. Information accessibility, limited relevant technical know-how, social and cultural biases, resource constraints, etc. are also additional factors that might impede the adoption of recommended agricultural technologies by disadvantaged groups with low SES such as small, resource-poor farmers, etc. (Adhikarya and Rogers, 1978).

As reported during the FAO's Global Consultation on Agricultural Extension, most extension services in both developed and developing countries are assisting large and/or commercial farmers. An FAO-sponsored survey of 207 agricultural extension organizations in 113 countries revealed that in 1988 approximately 58 percent of extension resources worldwide were directed toward commercial farmers, including specialized producers of cash crops and export commodities. Only 22 percent of extension resources were directed toward subsistence or small farmers (Swanson et.al., 1990). One of the consequences of such a situation is the widening of knowledge and benefits gap between large and small farmers. SEC's problem-solving orientation puts heavy emphasis on assisting small, resource-poor farmers and appears to be quite effective in undertaking special intervention programmes to reduce or narrow such a gap.

Furthermore, extension programmes do not always need to rely on agricultural research to provide "new" technology packages. There are usually many existing or "old" agricultural technologies or cultural practices which have not been properly practiced by the majority of farmers, especially by small, resource-poor farmers. Many KAP surveys conducted in developing countries revealed that plenty "old" agricultural technology packages are not known, accepted, and/or properly practiced by a large number of small farmers. For these target beneficiaries, special SEC intervention programmes focusing on "old/existing" technology recommendations or packages are most needed in order to narrow the agricultural knowledge and development benefits gap between large and small farmers (Adhikarya and Rogers, 1978).

5.5.
SEC Helps in Improving Extension Linkage with Training

For most agricultural technologies to be adopted and practiced properly by farmers, training for them on the applications or utilization of such technologies, especially through practical field-based instruction, hands-on demonstrations, etc., is needed. However, many studies and field experiences have shown that often farmers are not motivated and/or interested in attending or actively participating in training courses organized for them. Many reasons have been given for such poor attendance or participation in farmers training programmes, among others, lack of time, perceived irrelevance of a training course, unaware of the importance of the training topic, etc. However, one of the most important underlying causes for such a problem is the lack of a "felt-need" for participating in a given training programme among these farmers.

SEC results and experience indicated that by informing and motivating target beneficiaries, especially small, resource-poor farmers, on the importance of, and need for, adopting/practicing a given recommended agricultural technology, it also creates either a "perceived" or "felt" need among these farmers for more detailed and comprehensive information and clarifications, and thus training. As a result of an effective SEC programme, farmers' "demand" for training on various aspects of the campaign topics or recommendations can normally be expected. One of the lessons learned from SEC activities is that extension programmes, especially through SEC, can create farmers' awareness, motivation, and felt-need for training, and thus provide a conducive condition for effective and participatory-oriented farmers training. Hence, SEC is a critical precondition or prerequisite for improving the effectiveness and usefulness of farmers training programmes.

5.6.
SEC Reduces Extension System's Workload and Increases Its Coverage

One of the important features of SEC is the strategic and planned use of selected cost-effective multi-media channels and/or materials, through a combination of personal, group and mass communication approach or system. Unlike many other conventional extension approaches or programmes, SEC does not only rely on extension workers to undertake all or most of the extension tasks. At the FAO's Global Expert Consultation on Agricultural Extension in 1989, it was revealed that extension programmes in most countries, despite the limited transport facilities and vast areas to cover, had relied heavily on agricultural extension workers and other agricultural field personnel, with very little mass communication support. The extension agent : farmer coverage ratio in Latin America is 1 : 2,940, in Asia 1 : 2,661 and in Africa 1 : 1,809. It was reported that only 16 percent of the extension programmes worldwide utilized mass media/mass communication channels (FAO, 1990). Moreover, during the Consultation, it was also reported that an average of 26 percent of the extension workers' time was devoted to non-educational tasks (Swanson et.al., 1990). Thus, one of the weaknesses of many agricultural extension system or programmes, especially in developing countries, is the over-dependency on extension workers to undertake all sorts of agricultural development related tasks, — a significant portion of which are not educational or instructional in nature. As a result of such a problem, most extension workers are overworked, ineffective, and not able to have a wide outreach coverage.

As suggested in Fig. 2-5, for certain tasks that deal with informational and motivational activities, for instance, the involvement of extension workers may not need to be as intensive or active as if the tasks relate to educational, instructional, field-based training, or action-oriented activities. SEC experiences in many countries have demonstrated that by employing a multi-media strategy, extension workers' workload of non-educational matters can be reduced, so they can undertake more important and relevant tasks which can not be done as effectively by mass or mediated communication channels or materials.

The investment and operational cost of employing extension workers cost is one of the highest among agricultural extension expenditures. By properly utilizing such human resources to undertake the tasks they do best, SEC experience indicates that not only extension programme cost-efficiency can improve (i.e., in terms of coverage expansion or reaching more farmers per extension worker), but also its performance quality and cost-effectiveness can be increased (i.e., in terms of increasing the levels of farmers' knowledge, positive attitude, and proper practice).

5.7.
SEC Encourages Partnership with, and Participation of, Community-Based Organizations

In most SEC activities, community-based resources, including other non-agricultural institutions (schools, rural development agencies, mosques, churches, local government units, etc.), non-government organizations (cooperatives, peasants organizations, water-users associations, self-help groups, women's organization, etc.), and private sector (seed dealers, fertilizer suppliers, rural shops, etc.), as well as intermediaries such as teachers, school children, religious and community leaders, etc. have been actively involved to assist agricultural extension workers in campaign planning and implementation. Such a partnership and the "buying-in" by community-based resources into a given SEC programme can enhance and facilitate its sustainability and institutionalization (Maalouf, Contado, and Adhikarya, 1991).

An important lesson learned from the above-mentioned experience is that with participatory planning approaches, and close coordination procedures, within a given agricultural development policy and programme context, community support and resources can be mobilized to support planned extension activities. More importantly, the support and endorsement of influential community leaders and the involvement of community-based institutions provide legitimization and credibility to SEC objectives, activities, and messages. However, such a participation by relevant government and non-government organizations must be coordinated properly by the Ministry of Agriculture (i.e., agricultural extension service or department), and the collaborative activities should be consistent and compatible with existing agricultural development strategies, and in support of planned extension programmes. Training activities on the extension programme implementation details and its technical contents for non-agricultural personnel from the participating community-based organizations are also critical to the success of such a collaborative and participatory extension service.

5.8.
SEC Helps Revitalize Extension Workers' Professionalism

Involvement in a campaign activity can be considered as an additional burden by many extension workers who are already overworked with routine tasks. SEC experiences, however, suggest that if some types of incentives can be provided to these workers, their support and participation in campaign activities will be forthcoming. These incentives, however, do not need to be in a form of financial rewards. From SEC experiences in many developing countries, it appears to be sufficient to offer enhanced prestige or status symbols rewards in order to mobilize extension workers' support.

By providing extension workers with attractive and useful multi-media campaign materials which they can use with, or give to, their clientele (i.e., farmers), SEC programmes were able to boost the prestige and credibility of these workers among their friends and farmers. Previously, extension workers were perceived by many as having a low credibility. More importantly, SEC activities helped them gained improved self-confidence by providing new or additional know-how on the recommended campaign topics or messages through special training/briefing. In addition, all campaign media, including printed materials, radio and television programmes, etc., publicized the extension workers as being specially trained and thus highly qualified, and suggested that farmers contact them for further information and advice. Such positive image-building publicity conferred on the extension workers a certain status which served to increase their enthusiasm and motivation to be involved in extension campaign activities.

Another lesson learned from the SEC activities is that frequent field visits to sub-districts or villages by relatively high agricultural officials from provincial and/ or central levels to discuss and learn the problems and needs of extension workers and farmers are well appreciated. Such an effort can serve as a cost-effective means of boosting the morale and work enthusiasm of agricultural field personnel, and increasing the support and participation of community leaders and their members in agricultural extension an training activities.

5.9.
SEC Shows that Extension Programmes Can be Strategically Planned, Efficiently Managed, and Systematically Monitored & Evaluated

The SEC programmes undertaken in many developing countries have demonstrated that even with limited resources and utilizing existing facilities, agricultural extension intervention activities which have a programme focus with specific/measurable objectives can be undertaken successfully. The SEC process, including its conceptual framework and operational procedures/steps has increasingly been considered as a "microcosm" of how a routine agricultural extension system or programme can be planned, managed, monitored, and evaluated. Having been involved in SEC activities, and/or seen the actual achievements and results of such activities, many extension planners, managers and trainers, as well as their superiors and other agricultural decision-makers have taken initiatives of applying important SEC principles and methods for improving the agricultural extension programme.

One of the critical lessons generated from the SEC experience is that evaluation results should also include a process documentation detailing the important decision-making steps and/or procedures from the initial stage until the completion of the programme. Without such insights into the entire process, including the what, why, and how aspects, of the extension planning, training, management, and monitoring & evaluation activities, it would be unlikely that replications of an SEC programme can be done with a high degree of quality control.

5.10.
SEC Can Contribute in Improving and Strengthening Agricultural Extension Systems and Programmes

After more than a decade of experience with SEC programme implementations in many developing countries as listed in Fig. 4.1, including methodology development & testing, training materials preparation, and training of trainers activities, there are many important concepts, principles, and/or actions based on the SEC methodology which can be useful for improving and strengthening agricultural extension systems and programmes. Without elaborating on the contextual and operational requirements for applying these concepts or principles, as these are not the main focus of this chapter, and thus with a risk of over-simplifying them, some of most important concepts or principles are summarized below :

→ Problems identification and needs assessment should be undertaken using appropriate participatory approach and scientifically-sound methods, such as, micro-level baseline survey of farmers' Knowledge, Attitude, and Practice (KAP), including focus group interviews (FGI).

→ Extension objectives should be specific, measurable, problem-solving oriented, reflected in terms of knowledge, attitudes, and/or behavioural changes, and based on target beneficiaries' needs.

→ Strategic planning principles should be applied, especially in prioritizing problems to be solved by extension and training activities, and in audience analysis and target beneficiaries segmentation.

→ Resource allocation, including field personnel tasks assignment, should be based on cost-benefits and risk/payoff analyses and guided by specific extension programme objectives, strategy and plan.

→ Extension programmes must be planned, implemented and managed by a multi-disciplinary team of personnel, and this will require practical and workable functional linkages and collaboration among relevant agencies/staff dealing with agricultural research/technical subject matters, extension, training, communication support, etc. as well as concerned government and non-government organizations.

→ Participation and active involvement of relevant community-based organizations in support of specific extension activities under the coordination of a national extension service or agency should be sought. Community-based resources should be mobilized and utilized not only for the purpose of sharing the cost and burden of public extension service, but more importantly for facilitating the sustainability and institutionalization of some aspects of such a service.

→ Extension programmes or services should not only provide agricultural research agencies with feed-back from farmers about recommended technologies generated by research, but it should also provide feed-forward information on research needs of farmers. By providing such an "agenda-setting" service to agricultural researchers, extension workers may, as a result, obtain improved technology packages which are more relevant, appropriate, and useful to farmers' felt needs.

→ Application and utilization of a multi-media strategy for reaching segmented target audiences should be considered in planning any large-scale extension intervention programmes. Cost-effectiveness of the selected combination of personal, group and mass communication approaches should be carefully examined, based on the extension objectives and/or expected outcome of such a programme.

→ Formative evaluation, including pretesting of prototype multi-media support materials for use in agricultural extension and/or training programmes, should be conducted with a representative sample of the intended target audience for such materials. Reproduction of these materials should not be attempted before the necessary modifications based on the pretesting results have been completed.

→ Staff training, on both the technical subject matters/technology packages and specific extension training tasks, should be an integral part of the extension programme preparatory activities. Systematic training needs assessment and tasks analysis for staff training programmes should be conducted regularly to determine which new training subjects are to be offered. In many countries, it is not only training in new technical subjects that is needed. Training on platform and analytical skills also seems to be inadequate, such as on community and group organization, leadership and entrepreneurship development, participatory needs assessment methods, strategic planning and total-quality management principles, message design and multi-media strategy development, cost-benefits and risk/payoff analyses, management information system, new computer and communication technology applications, etc.

→ Management planning is an essential part of extension programme development, and such a plan should be used not only to guide in the field implementation or operations, but also as a basis for designing and conducting the management monitoring, supervision, and process evaluation activities.

→ Some extension programmes should be subjected to empirical studies based on quantitative data (e.g., from surveys) and qualitative information (e.g., from focus group interviews), to demonstrate the effectiveness, usefulness, and when possible also the cost-benefits of such programmes. One of the purposes of such an evaluation is to learn the strengths and weaknesses of such extension programmes so that further improvements can be made in future replications. The other, perhaps more important, purpose is to demonstrate the role, function, value, and benefits of well-planned and strategically designed extension programmes to influential agricultural policy and decision-makers, in order to secure more policy support and obtain increased allocation of resources.

References

Adhikarya, Ronny (1978). "Guideline Proposal for
a Communication Support Component in Transmigration Projects".
A consultancy report prepared for FAO/United Nations (project 6/INS/01/T), Rome, Italy : FAO.

Adhikarya, Ronny and Everett M. Rogers (1978). "Communication and Inequitable Development :
Narrowing the Socio-Economic Benefits Gap".
In *Media Asia*, Vol. 5, No. 1.

Adhikarya, Ronny and John Middleton (1979). *Communication Planning at the
Institutional Level : A Selected Annotated Bibliography.*
Honolulu, Hawaii, USA : East-West Communication Institute, the East-West Center.

Adhikarya, Ronny (1983). "A Planned Communication Support Strategy for
Increasing Ligation Acceptors in Bangladesh".
A consultancy report prepared for USAID, Dhaka, and the Asia Office of the International Project of
the Association for Voluntary Sterilization (IPAVS).

Adhikarya, Ronny and Heimo Posamentier (1987).
Motivating Farmers for Action: How Strategic Multi-Media Campaigns Can Help.
Eschborn, Frankfurt, Germany: GTZ (Deutsche Gesellschaft für Technische Zusammenarbeit).

Adhikarya, Ronny (1985). "Planning and Development of Rat Control Campaign
Objectives and Strategies for the State of Penang, Malaysia".
A summary report of a workshop on multi-media campaign planning for integrated pest control,
organized by FAO/United Nations (projects GCP/RAS/101/NET and GCP/RAS/092/AUL).

Boonlue, Thanavadee (1987). "Knowledge, Attitude, and Practice on
Pest Surveillance System in Chainat Province, Thailand".
Bangkok, Thailand: Chulalongkorn, University, October 1987.

Boonruang, Pote and P. Chunsakorn (1988). "Management Monitoring Survey of
the Pest Surveillance Campaign on Rice".
Bangkaen, Thailand : Katsetsart University, June 1988.

FAO (1990). *Report of the Global Consultation on Agricultural Extension.*
A proceeding of an international expert meeting held in Rome, 4 - 8 December 1989.
Rome, Italy : Food and Agriculture Organization (FAO) of the United Nations

Hamzah, Azimi and J. Hassan (1985). "Rice Farmers' Knowledge, Attitude and Practice of
Rat Control: A study conducted in Penang, Malaysia".
Serdang, Malaysia : Universiti Pertanian Malaysia (Agricultural University of Malaysia), August, 1985.

Hamzah, Azimi and E. Tamam (1987). "Rat Control Strategic Multi-Media Campaign for
the State of Penang, Malaysia : A report of the information recall and impact survey".
Serdang, Malaysia : Universiti Pertanian Malaysia (Agricultural University of Malaysia), August, 1987.

Ho, Nai-Kin (1994), "Integrated Weed Management of Rice in Malaysia: Some Aspects of the Muda
Irrigation Scheme". A paper presented at the FAO-CAB International Workshop on Appropriate Weed
Control in Southeast Asia, 17-18 May 1994, Kuala Lumpur, Malaysia.

Khor, Yoke Lim and Ramli Mohamed (1990). "The Information Recall and Impact Survey (IRIS) on the Strategic Extension Campaign on Integrated Weed Management in the Muda Irrigation Scheme, Malaysia".
Penang, Malaysia: Universiti Sains Malaysia (Science University of Malaysia), March 1990.

Khor, Yoke Lim (1989). "Strategic Extension Campaign Experiences on Integrated Weed Management in the Muda Irrigation Scheme".
Penang, Malaysia: Universiti Sains Malaysia (Science University of Malaysia), April 1989.

Maalouf, Wajih, Tito E. Contado, and Ronny Adhikarya (1991). "Extension Coverage and Resource Problems: the Need for Public-Private Cooperation".
In William M. Rivera and D.J. Gustafson (eds.) *Agricultural Extension: Worldwide Institutional Evolution and Forces for Change.*
Amsterdam, Holland : Elsevier Science Publishers B.V.

Middleton, John and Yvonne Hsu Lin (1975). "Planning Communication for Family Planning".
Honolulu, Hawaii, USA : East-West Communication Institute, the East-West Center.

Mohamed, Ramli and Yoke Lim Khor (1988). "Survey Report of Farmers' Knowledge, Attitude, and Practice (KAP) on Weed Management in the Muda Agricultural Development Authority (MADA)".
Penang, Malaysia: Universiti Sains Malaysia (Science University of Malaysia), March 1988.

Mohamed, Ramli (1989), "Strategic Extension Campaign Experiences on the Rat Control Campaign in Penang".
Penang, Malaysia : Universiti Sains Malaysia (Science University of Malaysia), April 1989.

Mohd. Noor, A.S. and Asna B. Othman (1992), "Strategic Extension Campaign (SEC) Training Programmes and Activities in Malaysia Since 1988", Telok Chengai, Malaysia: Extension Training and Development Centre (PLPP), Dept. of Agriculture.

Rivera, William M., A. Bannaga, and J. Phiri (1992). "Report of the Evaluation Mission - project ZAM/88/021 : Strengthening Agricultural Extension Services of the Department of Agriculture, Southern Province, Zambia".
New York, USA : United Nations Development Programme (UNDP).

Rogers, Everett M. and Ronny Adhikarya (1979).
"Diffusion of Innovations : An Up-To-Date Review and Commentary".
In Dan Nimmo (ed.) *Communication Yearbook 3.*
New Brunswick, New Jersey, USA : Transaction Books.

Rogers, Everett M. (1983). *Diffusion of Innovations (Third Edition).*
New York, USA : Free Press.

Swanson, Burton E., B.J. Farner and R. Bahal (1990). "The Current Status of Agricultural Extension Worldwide".
In *Report of the FAO's Global Consultation on Agricultural Extension.*
Rome, Italy: FAO of the United Nations.

Tiantong, C. (1988). "An Information Recall and Impact Survey (IRIS) on the Strategic Extension Campaign on Pest Surveillance System in Rice, in Chainat Province, Thailand".
A report prepared for Plant Protection Division, Dept. of Agricultural Extension, Bangkaen, Thailand, October 1988.

**Selected and Forthcoming FAO's Publications
on Agricultural Extension, Training, and Education.**

Swanson, Burton (ed.). *Agricultural Extension: A Reference Manual.*
Rome: Food and Agriculture Organization (FAO) of the United Nations, 1984.
(Also available in French, Spanish, Arabic, and Portuguese versions)

Oakley, Peter and C. Garforth. *Guide to Extension Training.*
Rome: Food and Agriculture Organization (FAO) of the United Nations, 1985
(Also available in French, Spanish, Arabic, and Portuguese versions)

FAO. *The Management of Agricultural Schools and Colleges.*
Rome: Food and Agriculture Organization (FAO) of the United Nations, 1985.
(Also available in French, Spanish, Arabic, and Portuguese versions)

Axinn, George. *Guide on Alternative Extension Approaches.*
Rome: Food and Agriculture Organization (FAO) of the United Nations, 1988.
(Also available in Spanish version)

FAO. *Report of the Global Consultation on Agricultural Extension.*
Rome: Food and Agriculture Organization (FAO) of the United Nations, 1990.
(Also available in French and Spanish versions)

Elliot, Sergio. *Distance Education Systems.*
Rome: Food and Agriculture Organization (FAO) of the United Nations, 1990.

FAO. *Make Learning Easier: A Guide for Improving Educational/Training Materials.*
Rome: Food and Agriculture Organization (FAO) of the United Nations, 1990.
(Also available in French and Spanish versions)

FAO. *Improving Training Quality: A Trainer's Guide to Evaluation.*
Rome: Food and Agriculture Organization (FAO) of the United Nations, 1991.
(Also available in French version).

FAO. *International Directory of Agricultural Extension Organizations.*
Rome: Food and Agriculture Organization (FAO) of the United Nations, 1991.

FAO. *Agricultural Extension and Farm Women in the 1980s.*
Rome: Food and Agriculture Organization (FAO) of the United Nations, 1993.

FAO. *Strategy Options for Higher Education in Agriculture: Expert Consultation Report.*
Rome: Food and Agriculture Organization (FAO) of the United Nations, 1993.
(Also available in Spanish version)

Wentling, Tim L. and R. M. Wentling. *Introduction to Microcomputer Technologies:
A Sourcebook of Possible Applications in Agricultural Extension, Education, and Training.*
Rome: Food and Agriculture Organization (FAO) of the United Nations, 1993.

Stringer, Roger and H. Carey.
Desktop Publishing: A New Tool for Agricultural Extension and Training.
Rome: Food and Agriculture Organization (FAO) of the United Nations, 1993.

Wentling, Tim. *Planning for Effective Training: A Guide to Curriculum Development.*
Rome: Food and Agriculture Organization (FAO) of the United Nations, 1994

Ausher, Reuben et.al.
The Potentials of Micro-Computers in Support of Agricultural Extension, Education, and Training.
Rome: Food and Agriculture Organization (FAO) of the United Nations, 1994

Forthcoming Publications:

Swanson, Burton (ed.). *Improving Agricultural Extension: A Reference Manual.*
Rome: Food and Agriculture Organization (FAO) of the United Nations.
(under preparation, expected out in 1995)

About the Author

Dr. Ronny ADHIKARYA has extensive professional working experience in the fields of education, training, communication, and extension, in support of agriculture/rural development and population. He has served for the Food and Agriculture Organization (FAO)/United Nations since 1981. Presently, he is a senior-level technical officer serving as Extension Education and Training Methodology Specialist at FAO's Agricultural Education and Extension Service in Rome, Italy.

He received his Ph.D. degree from Stanford University, California, USA, in 1981, with a specialization in communication and educational planning, a Master's degree (in development communication) from Cornell Univ. in 1972, and a B.A. (cum laude) in social and behavioural sciences from State University of New York in 1971. Prior to working for FAO, Dr. Adhikarya served in many developing countries as international consultant, for UNESCO, UNFPA, USAID, Ford Foundation, the East-West Center, and International Institute of Communication. He was a visiting professor at the Science University of Malaysia, a research staff at the East-West Center, Hawaii, USA and at Stanford University, Calif., USA, and also worked as an award-winning journalist in Indonesia. He has authored seven other books on non-formal education, extension and communication subjects, published in the USA, Germany, Italy, Singapore, and Malaysia, plus numerous chapters/articles in technical books and professional journals.

His present professional interests and activities are in the fields of strategic extension planning and management, multi-media learning methodology and materials development and testing, social marketing and behavioural change research, as well as information and computer technology applications for planning, training and learning within the context of environmentally-sound and sustainable agricultural development.

For more information on these publications, please write to:
Chief,
Agricultural Education and Extension Service (ESHE)
Human Resources, Institutions, and
Agrarian Reform (ESH) Division
FAO of the United Nations.
Via delle Terme di Caracalla
Rome 00100 - ITALY
Phone: (39 - 6) 5225 - 4001
Fax: (39 - 6) 5225 - 3152